Applied Modern Control

Edited by Le Anh Tuan

Published in London, United Kingdom

IntechOpen

Supporting open minds since 2005

Applied Modern Control
http://dx.doi.org/10.5772/intechopen.75856
Edited by Le Anh Tuan

Contributors
Espen Oland, Hideo Nakamura, Yoshihisa Saito, Ming-Sen Hu, Navin Khaneja, Rajeev Dohare, Natalia Bakhtadze, Vladimir Lototsky, Valery Pyatetsky, Tolochko Olga, Ana Elisabeta Oros Daraban, Popan Gheorghe, Jorge-Humberto Urrea-Quintero, Nicolás Muñoz-Galeano, Lina M. Gómez-Echavarría

First published in London, United Kingdom, 2019 by IntechOpen
IntechOpen is the global imprint of INTECHOPEN LIMITED, registered in England and Wales, registration number: 11086078, The Shard, 25th floor, 32 London Bridge Street
London, SE19SG – United Kingdom
Printed in Croatia

British Library Cataloguing-in-Publication Data
A catalogue record for this book is available from the British Library

Additional hard copies can be obtained from orders@intechopen.com

Applied Modern Control
Edited by Le Anh Tuan
p. cm.
Print ISBN 978-1-78984-826-7
Online ISBN 978-1-78984-827-4

For EU product safety concerns:
IN TECH d.o.o., Prolaz Marije Krucifikse Kozulić 3, 51000 Rijeka, Croatia, info@intechopen.com or visit our website at intechopen.com.

Meet the editor

Le Anh Tuan graduated with his B. Eng. and M. Eng. in Mechanical Engineering and Marine Machinery from Vietnam Maritime University in 2003 and 2007, respectively. He received his Ph.D. degree in Mechanical Engineering from Kyung Hee University, South Korea in 2012. He was a Research Fellow at Nanyang Technological University, Singapore, a Visiting Scholar at the University of Birmingham, UK, and an Endeavor Research Fellow at the University of Technology Sydney, Australia. He has been a Contract Research Professor at the Center of Wind Energy System, Kunsan National University, Korea since 2016. He is currently an Associate Professor in the Automotive Engineering Department of Vietnam Maritime University. His interests include nonlinear robust adaptive control, dynamics, vibration, and control of industrial machines.

Contents

Preface

With this book, we introduce to the readers a monograph entitled *Applied Modern Control*. This book introduces recent studies on control and automation using modern techniques. The book contents are represented in two sections:

Section 1 introduces the modern control systems from the large complex systems such as train operation systems to micro and nano control systems. Section 1 composes of four chapters: Chapter 1 reviews the smart control system for train operation with the application of Internet of things and Safety 2.0. Chapter 2 discusses a general picture of flexible control and measurement in nanotechnology. Chapter 3 proposes a practical instrument and control system of glove box for high-vacuum deoxygenation and water removal. Finally, Chapter 4 synthesizes a control system for dividing wall columns on the basis of artificial neural networks.

With five chapters, Section 2 applies control and identification techniques to practical plants such as satellites, power electronic converters, and interior permanent magnet synchronous motors. Chapter 5 represents a dynamics model and attitude control of satellites in elliptical orbits. Chapter 6 presents system identification algorithms using associative search of analogs and wavelet analysis. Chapter 7 proposes the control algorithm for power electronic converters on the basis of zeros placement techniques. Chapter 8 presents the optimal control solution for interior permanent magnet synchronous motors using “Maximum moment per ampere” and “Maximum torque per volt” principles. Lastly, Chapter 9 describes time-optimal control of coupled spin dynamics applied in spectroscopy of nuclear magnetic resonance, quantum computation, and information.

As the editor, I would like to thank all the authors who contributed their excellent studies for the book. Hopefully, the readers will find much useful information in this book.

Le Anh Tuan
Associate Professor at Automotive Engineering Department,
Vietnam Maritime University,
Hai Phong, Vietnam,
Research Professor at Center of Wind Energy System,
Kunsan National University,
Gunsan, South Korea

Section 1

Modern Control Systems

Chapter 1

Study on a New Train Control System in the IoT Era: From the Viewpoint of Safety2.0

Hideo Nakamura and Yoshihisa Saito

Abstract

Safety2.0 which advocates cooperative safety is attracting attention. Assuming that Industry4.0 proposed by the German authorities is an IoT-based production revolution, Safety2.0 is a Japanese-originated proposal that seeks to create a more flexible and sophisticated safety by introducing Internet of Things (IoT) into production sites. This chapter introduces the concepts of Safety2.0 and its spread internationally, focusing on the activities of IGSAP, a Safety2.0 promoter. Furthermore, we look back on the conventional train control from the viewpoint of Safetyx.x and look at the appearance of the train control suitable for Safety2.0 using IoT. As a result, in this chapter, we propose a simple and smart train control system unified train control system (UTCS), in which a train control system is realized in a hierarchical structure of a logic layer, a network layer, and a terminal layer, and discuss its processing method.

Keywords: Industry4.0, IoT, Safety2.0, UTCS, ATP-block system, train control system

1. Introduction

The safety of manufacturing today is mainly based on European-style methodology mainly based on certification found in functional safety standards (IEC61508) and the like. Since stakeholders have different standing positions for certification, there is a reflection that productivity cannot be improved by mutually balancing each other. Safety2.0 focuses on the realization of "cooperative safety" that creates a highly safe condition by mutually exchanging information among the elements that constitute systems based on IoT. Therefore, under Safety2.0, stakeholders are required to share their wisdom with each other and to manufacture products based on IoT. What is the meaning of "relying on IoT" for railways? Radio train control systems advanced train administration and communications system (ATACS) and communications-based train control (CBTC) will be analyzed and evaluated from the viewpoint of Safety2.0, and future train control will be considered.

2. Concept and current state of Safety2. 0

The new safety concept named Safety2.0 is not unrelated to the desire to overcome occupational safety occlusions at production sites. In the summer of 2015, the Safety2.0 Preparatory Committee (renamed the "Safety2.0 Promotion Committee" in 2016) was established. The conclusion was that the essential elements of the system exchanged information with each other to create optimal safety, which was the construction of a cooperative safety methodology suitable for the IoT era [1, 2]. Safety2.0 was, of course, preceded by Safety0.0 and Safety1.0, which supported present-day safety. However, the use of IoT is very effective in overcoming the sense of occlusion on it. In this chapter, we review the changes in safety initiatives from the perspective of Safety0.0 and Safety1.0, and confirm the today's status of Safety2.0. In addition, practical activities aimed at realizing Safety2.0 have begun, and we would like to introduce the situation.

2.1 Safety0.0 to maintain safety with sustained arousal

Direction calls are famous for railway safety culture. This is also an easy-to-understand case of Safety0.0, which attempts to prevent accidents by drawing attention and keeping the spirit awake at all times. Japan has been regarded as a leader in the sustainable implementation of the "Zero-Accident Motion" and other activities leading to Safety0.0. Looking back at this, various activities have been carried out, including the enactment of the Occupational Safety and Health Law in 1972 and the start of the "Everyone's Participation in Zero Accidents" motion by the Center Industrial Accident Prevention Association in 1973. On the other hand, it is interesting to note that the European version of the Zero-Disaster Motion, which is a top-down movement but corresponds to Safety0.0, is beginning to take place in Europe, where the International Society for Social Security (ISSA; International Social Security Association) has achieved outcomes at Safety1.0 through the launch of Vision Zero and the launch of Zero Accidents Forum by Europe and Finland.

2.2 Safety1.0 to prevent accidents

There are limitations to ensuring safety by Safety0.0 alone, since no mistakes will be made and the machinery will be destroyed, no matter how well people are trained. Therefore, technological efforts to implement some sort of safety measures for "goods" such as machinery and systems have progressed on the assumption that "people make mistakes" and "machinery breaks down." In Europe, this was established as a mandatory standard, and work accidents were greatly reduced by providing industrial machinery with safety protection measures. The basic idea was to establish a barrier between industrial machinery, which is a hazardous source, and humans, and to establish a mechanism for moving machinery only when humans are absent. This is Safety1.0. In Japan, the Industrial Safety and Health Law was revised in 2005, and risk assessment was added as an obligation to make initiatives. Efforts learned from European experiences and outcomes are now being developed. For this implementation, a number of safety mechanisms have been developed and incorporated into devices at each industrial site. However, even though it is isolated from hazardous sources, it is not possible to completely isolate workers during maintenance. In addition, many sites are difficult to isolate, such as construction work.

2.3 IoT-based Safety2.0 concepts

Safety2.0 is a system that ensures methodology suitable for the IoT era, in which the essential elements constituting the system exchange information with each other to achieve optimal safety [1, 2]. In this respect, it is different from the idea of relying on human attention (Safety0.0) and of taking some protective measures against human errors and mechanical failures to ensure safety (Safety1.0). These relationships are summarized in **Figure 1**. In Safety0.0, risks exist in a wide range, including

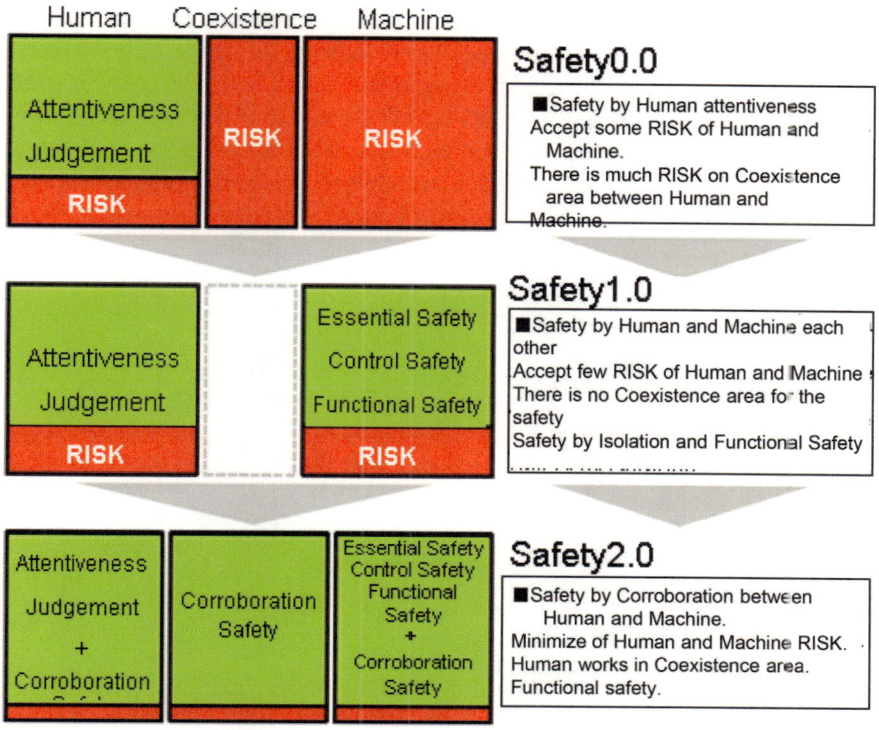

Figure 1.
Comparison among Safety(x.x).

coexistence areas, in order to prevent accidents due to attention and judgment. On the other hand, in Safety1.0, the risks were reduced by dividing the human area into the machinery domain and by providing various safety measures in the machine domain. Safety of isolation that creates as little coexistence as possible is fundamental.

On the other hand, in Safety2.0, machineries and humans exchange information and cooperate with each other to ensure safety, so that both machines and humans can coexist with each other. In addition, since appropriate safety using information is maintained during operation, the overall risk is greatly reduced.

The Nikkei BP brochure [1] explains Safety2.0 as follows: "Frankly speaking Safety2.0 is a collaborative safety built by people, goods, and the environmental in cooperation with each other. The best mean to achieve this Safety2.0 is to make rapid progress today. In Safety1.0, there was only a choice between "stop" and "go." In Safety2.0, however, detailed operations are carried out by exchanging information between people and machineries, and the safely coexistence between both is aimed at (Omitted hereafter.)." A case of Safety2.0 is indicated in **Figures 2** and **3**.

Management understandings and support are essential to Safety2.0's in-house promotion. Fortunately, IoT-based safety technologies were widely developed prior to the conceptual development of Safety2.0, and there were certain grounds to accept Safety2.0. However, in order to expand Safety2.0 from Japan to a wide range of countries as well as in Japan, we would like to have a promotion base. To this end, the Safety Global Promotion Mechanism (IGSAP; The Institute of Global Safety Promotion) was established.

2.4 Current stage of Safety2.0 (activities centered on IGSAP)

IGSAP has established the safety management forum as a forum for managers and managers to gather and replace information and experiences in order to actively engage in the safety of customers, employees, and the safety of the company as a company. In addition, Japan has been vigorously inviting European opinion leaders to work to eradicate occupational accidents under the Vision Zero and replace opinions through visits to Europe. In February 2018, NIPPO's automatic stop tire roller/wheel loader, equipped with an emergency stop technology for construction machinery, was the first Safety2.0 automatic stop tire roller/wheel loader to be registered.

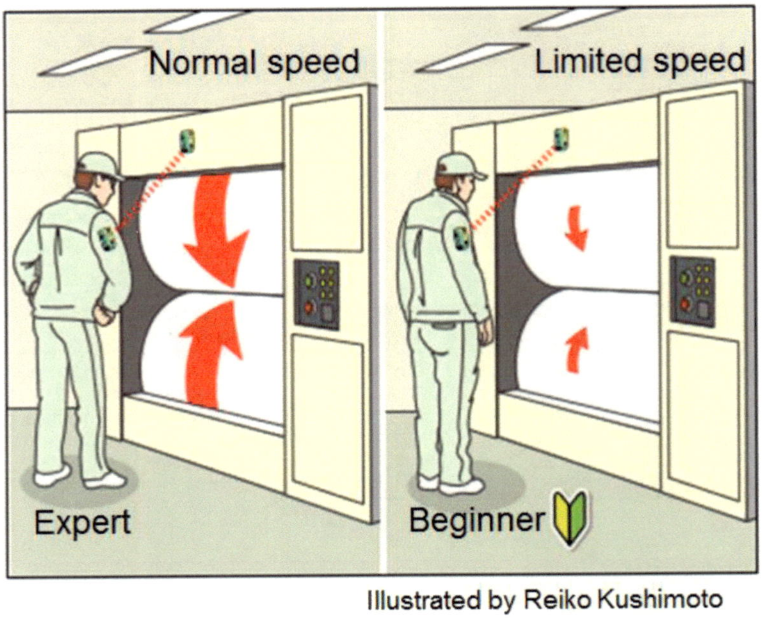

Figure 2.
An example of Safety2.0 (Case 1).

Figure 3.
An example of Safety2.0 (Case 2).

3. Outlook for train control systems from the viewpoint of Safety2.0

On October 14, 1872, Japan's railway opened between Shimbashi and Yokohama. Following the opening of the railway, the railway construction regulations and railway dormitory train transport regulations were enacted, and other rules for ensuring safety operation were rapidly established. This is a summary of the standards and regulations that form the base of Safety0.0. Technologies have also been imported overseas. For example, 1887, a voucher-type blocking method with the use of occlusion telegraphs was established between Kyoto and Osaka, and a second-class mechanical interlocking device was installed at the junction of the Yamanote Line and the Tokaido Line in Shinagawa Station. In addition, the manufacture of railway signals and interlocking equipment began at the Mimura Factory of Tokyo Tsukishima, and the move toward in-house production also progressed [3].

Signaling devices based on these technologies are mainly designed to ensure routes and link signals, and the safety of drivers can be guaranteed if they operate according to the signals. However, there was an accident caused by a mistake by the station manager that the train line between stations was forgotten and the opposite route was set in the single-line section. The emphasis was placed on raising human attention in the meaning that mistakes could lead to accidents as they were.

3.1 Safety1.0 and train control

3.1.1 Actual state of safety assurance based on Safety1.0

Machinery is a hazard source in the world of machine safety, such as factories. For this reason, a mechanism for confirming that no human or body part exists in the work area of the machinery and allowing the machine to operate only at that time has been adopted. Furthermore, the sensors used therein and the devices for confirming safety are fail-safe, and the Safety1.0 mode is constructed.

On the other hand, trains are the main source of hazard in railways. For this reason, the "concept of isolation" that allows only one train to exist in one section was established as a blockage. In addition, in the station premises, an interlocking device has been developed, which ensures absolute safety even if erroneous signal handling is performed. In addition, the use of fail-safe orbit circuits has resulted in advanced signal systems. However, even under these mechanisms, mistakes of the driver, such as signal advancement, can cause an accident. For this reason, in-vehicle alarms have been developed, which give an alarm to wake up when a stop signal is approached, and they have evolved to an ATS, which applies an emergency brake when a driver does not perform a predetermined treatment.

However, the technology introduced in the field of railways and industrial machinery that is safe but that is basically safe to stop was not the same as the technology introduced in the case where flight continuity is safe instead of stopping as in the case of an aircraft. In addition, own technologies were developed in the industrial machinery field and railways. As a result, no agreement was reached on common safety and fail-safe technologies across industries, and there was no common measure of safety.

3.1.2 Computer-based safety control device and Safety1.0

In the 1980s, computers were introduced into Safety1.0, which had been secured with sophisticated circuitry. As a result, the aspects of safety technologies that have been uniquely pursued in each industry have changed. In addition to the conventional technology, the concern of engineers in various industries has been to ensure the safety of the computer itself (hardware) used and to prevent bugs (quality assurance) in the software to be incorporated.

Computer hardware safety has been solved by redundant configuration and verification of processing results and appropriate integration of diagnostic circuitry, but the methodology has been discussed across industries. Software has a common issue: how to develop high-quality, bug-free software. What is important is that the sophisticated circuits that have once been inherited as the essence of safety technology by various industry sectors have all been incorporated as software logic and have not appeared on the surface.

This resulted in a deeper recognition of common methodologies across industrials and the establishment of a new IEC61508 of international standards that can be applied across industries under the new concept of "functional safety." In order to ensure safety in the age of functional safety, first of all, the safety level of the target systems is determined as safety integrity levels (SILs) as a result of the risk analysis. In addition, the design requirements of the hardware and the targets of the hazardous side failure rate are indicated in accordance with the SIL value. In software, the design requirements for each phase of the life cycle are determined according to the SIL value. Thus, "risk" became a common measure of safety. On the other hand, the authenticating organization has evaluated the validation of the determination of the SIL value and the validity of the specific work according to the value of the SIL.

The situation is different in the age of reliance on circuit technology and in the age of use of computers. Nevertheless, the concept of Safety1.0, which seeks to safeguard safety by protection measures in the event of human error or device defect, is common.

What is important is the fact that computer use is evolving the control system into a more sophisticated one. There is a great difference between the age and today of the development of computerized signaling devices that have solved the issue of how to make hardware and software safe and have replaced safety technology with program logic to produce electronic interlocks and level crossings.

The elimination of concerns about the use of computers for safety control has facilitated the addition of advanced functions. In addition, the successful use of diagnostic technology has opened the way to integrate communication technologies such as networks and wirelesses into safety control devices. The issue of "train control system using IoT" is also due to the fact that communication including wireless communication can be freely used for safety control. Today, the challenge of developing new computer-based systems is continuing, and sophisticated control systems are emerging. In 2011, the world-first full-fledged wireless-train control systems ATACS was launched on the Sensei Line and has been achieving excellent results. Based on this achievement, it was also introduced between Ikebukuro and Omiya on the Saikyo Line in 2016, and has been operating stably. The SPARCS (simple-structure and high-performance ATC by wireless communication system) of radio train control systems developed by Nippon Signal Co. Ltd. is also well received oversea. What are the relationships between these advanced systems and Safety2.0? What kind of system should we look at next to these advanced systems? We would like to consider on the base of the specifications of the computer- and radio-aided train (CARAT) control system in which the author participated in the development when he was in the Railway Research Institute.

3.2 Advanced train control systems and Safety2.0

The interlocking of the CARAT is called point-control. The position of the train is managed by the block-ID and the position (kilometer) in the block. A plurality of trains can exist in one block except for the section of the point machine in order to realize the movement blocking even in the station premises. Point control does not include path locking or segmented locking. Since the route is pulled back by exchanging information with the on-board device, it is not determined only by the train position. A reasonable and safe process is substituted for the time of the access lock, which was uniformly applied when the line is in the approaching section. The prototype of this point control was installed at Tsubame-Sanjō Station, and the function was confirmed by a monitor run.

Since the CARAT was designed to cope with the Shinkansen, the level crossing control function is not required. However, investigation of functions and check of the effect by simulation were carried out, and it was proven to be effective for the fixed-time control and safety improvement, which had been regarded as an issue of the existing level crossing. As a result, the point control and the level crossing control are positioned as processes for extending the point where the train is allowed to travel. The interval control device generates the "traveling permit point information" and transmits the result to the on-board device, thereby making it possible to unify the processing in the interim of the station and in the premises of the station. The form in which each of the emerging devices performs reasonable processing while exchanging information in various directional is precisely located in the IoT. The compatibility between CBTC and Safety2.0 appears to be good.

3.3 Outlook for future train control systems

Existing train control systems have condensed know-how learned from the experience of large accidents caused by human error. As shown in **Figure 4**, the basic control function is the blocking function and the interlocking function. However, safety cannot be ensured by this function alone. Today's safety is achieved in cooperation with safety devices such as ATS and ATC for the objective of preventing accidents caused by human error. Nevertheless, as shown in **Figure 4**, the actual situation is a complex combination.

A simple and orderly system as shown in **Figure 5** emerges from the IoT-based train control system. On the scene, there are only point machines, level crossings, and trains that make up the route of travel. The processing unit of the center directly transmits "information up to the section where the vehicle can safely travel" to the on-board device of the train as a control command (travel command) In the CARAT, the interval control device sent the "travel permission point information" to the company office device below the information of the point control. The interval control device is centralized at the center, and at the same time, both the point control and the level crossing control are centralized at the center. Therefore, this form is organized only by adding the IoT viewpoint to that demonstrated in the CARAT.

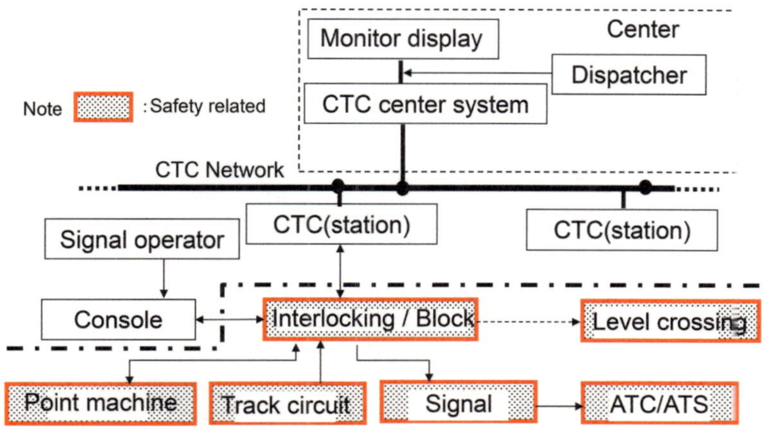

Figure 4.
An architecture of conventional train control system.

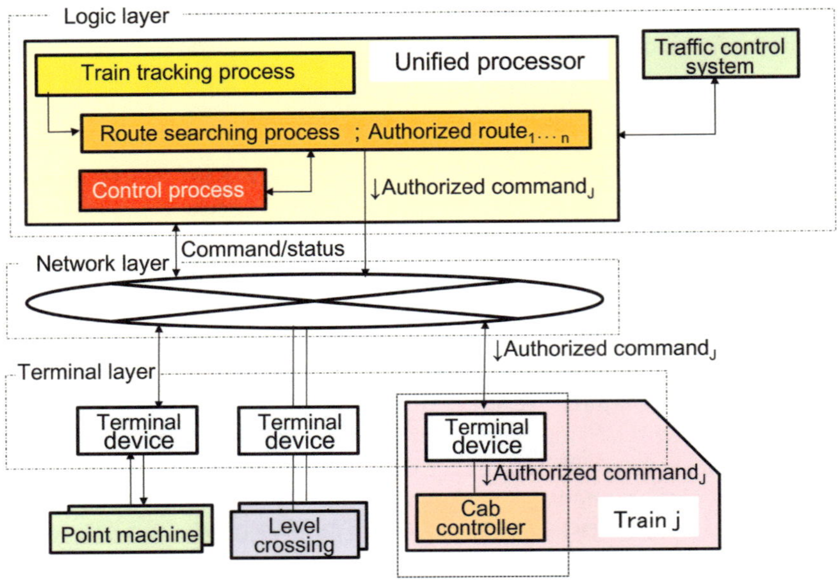

Figure 5.
System architecture of unified train control system.

The "interlocking device," "blocking device," and "ATC/ATS," which have been so popular in the signal field as to be the three types of gods, disappear. However, it is not the introduction of the "centralized linkage device." Existing interlocks have incorporated various locking logics to ensure the safety of whatever handling is done by the signal handler. In this respect, the centralized interlocking device does not change at all. Instead, it is claiming to replace the complex interlocking logic itself with point control, which makes it unnecessary. Point-control algorithms have been demonstrated in the monitoring run of the next-generation train control systems CARAT carried out by the Railway Institute on the Joetsu Shinkansen.

The overall system architecture consists of a terminal layer and a center device (functional layer) for controlling site equipment/on-board safety control devices, and an IP network (network layer) connecting them. This level next-generation system is named unified train control system (UTCS), and various studies are being conducted [5, 6]. This system also conforms to the concept of Safety2.0, which consists essentially only of the equipment necessary for the system: trains, point machines, level crossings, and center equipment, and "the essentially necessary equipment exchanges information with each other and realizes functions (I have called this intrinsic control)."

3.4 Outline of the UTCS processes

In the UTCS, the concept of a "path" (labeled "authorized route" in **Figure 5**) for a train is introduced for the standardization of processes. A path means a "limit position to which running is possible," and is derived from an associated preceding train, a point machine, and the states of a level crossing for each train. For this reason, train processes by unified processors are realized by train tracking, path searching (or "route searching"), and control processes that are initiated by route searching processes in order to control level crossings and point machines.

When paths for trains are determined, an authorized command with additional speed restriction information in a path is also generated and sent to the corresponding terminal device of the terminal layer. Path searching creates a search for a limit point to which running is possible (a path) in the train movement direction. In the case of station premises, however, a search is made according to a scheduled running path acquired from the running control device (or "traffic control system") on the functional layer. The path at this time is based on the terminal end of the running path and is determined by the state of point machines existing in between and at the tail position of a possible preceding train (including the safety margin).

On the other hand, in the case of a midway point between stations, the tail position of the preceding train or the state (labeled "status" in **Figure 5**) of an existing level crossing is associated with the determination of a path. If the level crossing is controlled by the relevant train and the status indicates "passing allowed," which means closing completion and no obstacle, the search is extended up to a further remote position. Although on-board devices are responsible for on-board safety processing, a continuous speed check according to a pattern is realized on the train anyway. Moreover, in the case of the CBTC, a high-level speed check function can be realized by installing a terminal device on the train, rather than providing an ATP terminal device on the ground. In **Figure 6**, a data flow under UTCS is illustrated.

An example of a UTCS that relies on Safety2.0 is ATP-block system [4–6], which is typical of intrinsic control, although the detail of the ATP-block system is omitted, and

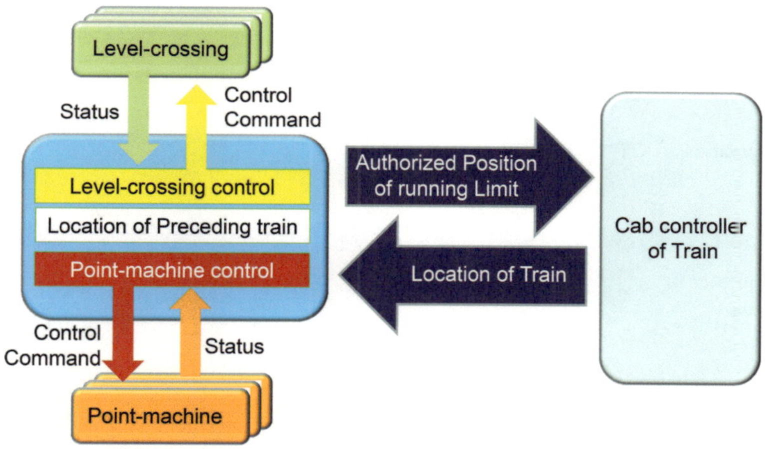

Figure 6.
Data flow under UTCS.

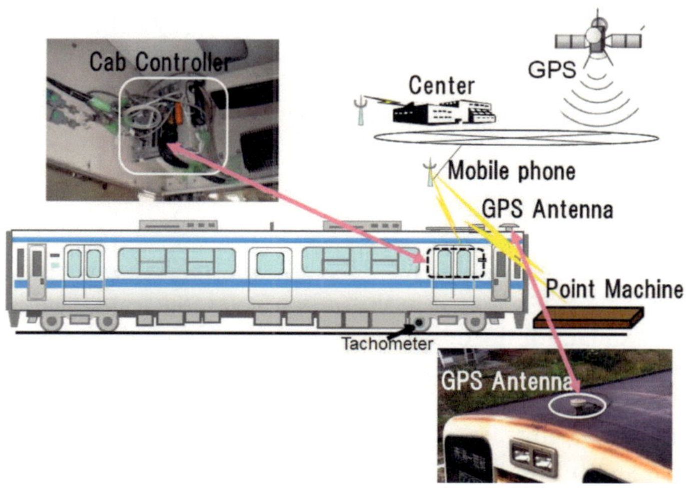

Figure 7.
Components of ATP-block system.

the essential devices constituting the system mutually exchange information as IoTs, thereby realizing advanced functions. In the case of ATP-block system, the blocking device and the interlocking device, which were previously located at the station and controlling the driving direction with the adjacent station, disappear(see **Figures** 7, 8) The history of technological progress in a train control system and its relation to the Safety(x,x) is illustrated by **Figure 9**.

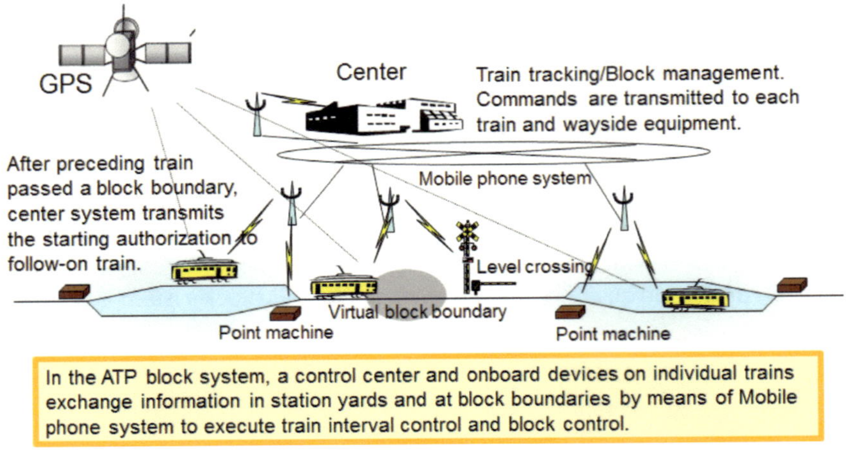

Figure 8.
System configuration of ATP-block system.

Figure 9.
History of technological progress in a train control system and Safety(x.x).

4. Afterword and conclusion

Development of new systems involves certification work in accordance with international standards. Especially in the train control, the train is subjected to the baptism of the standard of the reliability, availability, maintainability and safety (RAMS; IEC62278). To make this baptization smart, it avoids complications and makes the system as simple as possible. Railways are one of the social systems and have a long service life. Erroneous selects can leave the roots of the trouble. Examining and simplifying the components as much as possible improves the system's visibility and facilitates certification. Furthermore, the reliability is increased by the reduction of the amount of goods, and protection becomes unnecessary. The fewer the number of interfaces, the greater the safety. In addition to the advantages of these nonfunctional requirements, we believe that systematization in accordance with Safety2.0 can be an informative methodology in this regard. Under these circumstances, if information such as orbit protection is automatically extracted by AI based on vehicle vibration data and the like during daily driving, the railway can be reconstructed as a competitive transportation means.

Safety2.0 is an initiative that contributes not only to safety effects but also to productivity improvements and contributes to management. I hope that UTCS will be a successful development. We believe that this will also contribute to the dissemination of Japanese Safety2.0.

Author details

Hideo Nakamura[1*] and Yoshihisa Saito[2]

1 Visiting Researcher of the University of Tokyo, Japan

2 Kyosan Electric Mfg. Co., Ltd., Yokohama-Shi, Kanagawa, Japan

*Address all correspondence to: nakamura.hideo@nihon-u.ac.jp

References

[1] Concept Safety2.0. Safety2.0 Projects of Nikkei BP. Tokyo; 2016

[2] Nakamura H. [Invited Lecture] New safety culture and manufacturing on Safety 2.0. IEICEJ Tech. Report of DC-Research Group. Vol. 116, Issue 373; 2016. pp. 47-50

[3] History of Japanese Railway Signal. Tokyo: Signal Safety Association of Japan; 1980

[4] Saitou Y, Asano A, Nakamura H, Mochizuki H, Takahashi S. Restructuring of train control system by hierarchical design. IEEJ Transactions on Electronics, Information and Systems. 2016;**136**(7):923-928

[5] Saito Y, Asano A, Nakamura H, Takahashi S. A proposal for the design of integrated train control systems capable of improving reliability and safety. In: Proceedings of the Third International Conference on Railway Technology: Research, Development and Maintenance; Civil-Comp Press, Paper70; 2016. pp. 1-11

[6] Nakamura H. How to deal with revolutions in train control systems. Engineering. 2016;**2**:380-386

Chapter 2

Flexible Control in Nanometrology

Gheorghe Popan and Ana Elisabeta Oros Daraban

Abstract

The unceasing development of new small products has increased constantly by introducing multiple facilities in line production, reduced life cycles of new innovative products, and high-precision techniques that require automation and robotization of the nanotechnology production processes. Classic size products are made in normal series and deal little change over the years, while in the field of nanotechnology, product life cycles were shortened significantly, and series production must adapt to the market challenges. Considering the fast changes and multiple innovations in production, we propose equipment that offers a high degree of flexibility and performance for quality products. To compensate efficiently, the fluctuations may appear in production series; a flexible control system is designed to adjust production for large number of items or for various models of processing. The control equipment dedicated to nanotechnologies developed by INCDMTM Bucharest offers solutions for automation processes adapted to various operations and for quick response occurring in nano-production. A modular special design offers flexibility during the process, handling and interoperable ones, along with the possibility of changes facilitated by software that controls the entire verification process and parameter selection for each checked item's admissibility.

Keywords: measurement system, calibrating nanotechnology, nanometrology, optical measuring system, laser measurement system, AFM control

1. Introduction and research problems

The rapid evolution in the field of technologies related to nanomanufacturing and nano-devices based on electrical, optic, magnetic, mechanic, chemical, and biological effects would allow measurements in specific length ranges involved. Moreover, the spectacular development of nanotechnology in recent years generated the development of new devices and smaller components, trends that have created the need to measure them by developing a new nanometrology field. For standard products, measurement and control systems and equipment have been created in hundreds of years, but for nano-metric components, new appropriate measurement systems must to be created quickly. In most cases the physical principle used to measure in the usual nano-production flow from a technical point of view does not correspond with normal measurement systems. Traditional measuring means have proved some technological limits in terms of accuracy because of the physical law constraints [1, 2, 3]. Furthermore, microsensors, transducers, and ultra-accurate machines must be calibrated or verified during production and, afterward, before reception at beneficiary, because it is through them that the measuring unit is transmitted to dedicated users, meaning final producers [4].

Control and measurement techniques in nanotechnologies face specific challenges at the actual incipient stage and form tolerances of the nano-products exceeding actual measurement equipments and standards, and new generation of performant electromechanical systems is required in the field of nanometrics [5, 6]. Thus, innovative devices based on new measurement principles have been used and developed. Industrial production implies increasing manufacturing speeds on the one hand and increasing accuracy of manufacturing on the other. This can be achieved by automating and robotizing both production and production control.

Different industries developed new innovative products or materials involved that currently utilize nanotechnology. The nanoscale analysis of biosystems and of specific materials started years ago (beginning of the twentieth century) when chemistry and physics allowed small-scale characterization (bacteria, fungi studies). Recent development of medicine applications, nano-characteristics of drugs or nano-surgery, has generated advanced progress in engineering building new nanoscale systems and creating new nano-technics [7].

Other areas of emerging technologies include semiconductors and optoelectronic design and production, which increase the progress of information and communication technologies (ICT). More and more positive results engaged new initiatives and contributed to develop nanotechnology applications for structures smaller than 100 nm. Actual growth of semiconductor industry exploded toward nanotechnology boost and industrial demand raised in the last few years, generating unstable economic expansion for electronic devices in term of quality.

New nanotechnologies penetrated globally in large areas, from electronics to optical devices and from new materials to biological systems, considering upgrade of specific and customized makers offering optimal and functional parameters of the new products. This is further relevant conceiving nano-systems based on optical, electronical, mechanical, and biological nano-devices [8].

In Romania and widely, we only find significant research and innovation projects for nano-systems and nanometrology reaching the TLR 3–4 level, stage that needs upscaling to TLR 6–8. Further, industrial nano-production needs calibrated applications and metrological infrastructure at nano-dimensions to be scaled up from laboratory stage to industrial systems, which follow the quality parameters of the production flow for every relevant process [9].

Evolving toward precise production, innovations are required for efficient production structure of control systems by designing them for accreditation; thus, some procedures ask for specific parameters that are necessary to be checked.

Nanoscale dimensional accuracy covers a narrow range of tolerances. Industrial systems in nano-production can't detect smaller deviations beyond the normal tolerances, and that may have unpleasant effects by damaging the production systems. Any nano-production system requires rigorous control and verification procedure based on dimensional checking; the field of nanometrology is not developed accordingly [10].

Research and innovation in nanometrology expand the number of interested scholars, who will be supporting widely new sustainable production of nano-devices, nano-systems, and nano-materials.

Industrial processes, from medical devices to aeronautics, involve a structure where process accuracy and product quality are supervised by a system of characteristic control for every landmark product, ensuring interchangeability of product parts and the functional parameters of the product [11].

Only a few organizations have integrated this kind of research; most applications are limited at laboratory findings. The main barrier of using nanotechnology control at large industrial scale is the lack of specific infrastructure; for that reason this study proposes some solutions (**Figure 1**) [8–10].

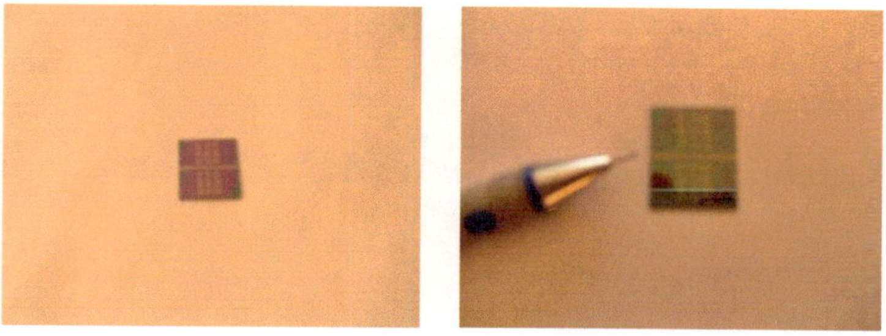

Figure 1.
Nano-electronics devices.

The equipment for nanometrology further presented is based on the experience of more than 30 years in research and didactic activity of the main author in the field of measuring devices and dimensional control systems. Activity in the field began with the design of control systems and devices from precision mechanics, with laser measuring and controlling equipment (laser probe heads, laser beam-scanning measuring systems, 3D cordless measuring machines, laser head, laser camshaft measuring equipment, laser calibration of coordinate measuring machines, and precision tool machines) [12].

The advantages of studying and realizing these systems are the basis for the next research and innovative solutions for nano-industries in terms of quality and precise manufacturing. The disadvantages are that systems and benchmarks to be verified in nanotechnological production are not palpable and in most cases are easily deformable and only distinguishable by a microscope. The transition from metrology to nanometrology required a new approach. Touch contact systems can no longer be used; the appropriate optical measurement principle—video inspection, laser scanning, and atomic force microscope (AFM) testing—must be approached.

The equipment is designed and developed by a multidisciplinary team from INCDMTM (National Institute for Research Development for Mechatronics and Measurement Technique) in Bucharest. The equipment is mainly driven by the need to control the production flow of a recognized mobile phone company based in Romania. Meanwhile, the mobile phone company ended its tax-free period and relocated its production from Romania to another country.

The chapter is structured according to primarily an introduction which highlighted the state of the art of this theme and secondly to describing the main parts of the equipment structure (experimental model). This includes Subchapter 2.1 that shows the optoelectronic control system including the charge-coupled device (CCD) camera, with examples of controllable nano-sensors; Subchapter 2.2 presents briefly the following control station with the laser control system; and then in Subchapter 2.3, the control station with the atomic force microscope (AFM) is shown, followed by a brief conclusion and direction for future research.

2. Experimental model

The experimental model presented in this paper for an innovative control and calibration equipment is built based on rotary feeding systems including table supports which are installed very precisely holding the nano-devices that need to be verified and calibrated. The equipment design allows calibration for a series of electronic nano-devices, bio-nano-devices, nano-materials, nano-sensors, or other nano-devices (**Figure 2**).

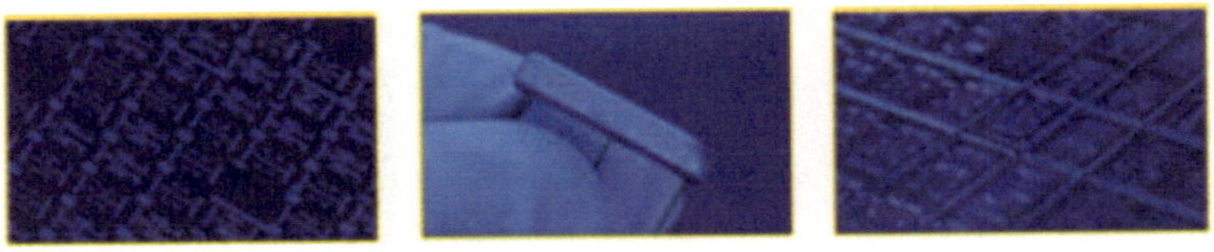

Figure 2.
Variety of Nano-devices necessary to be verified.

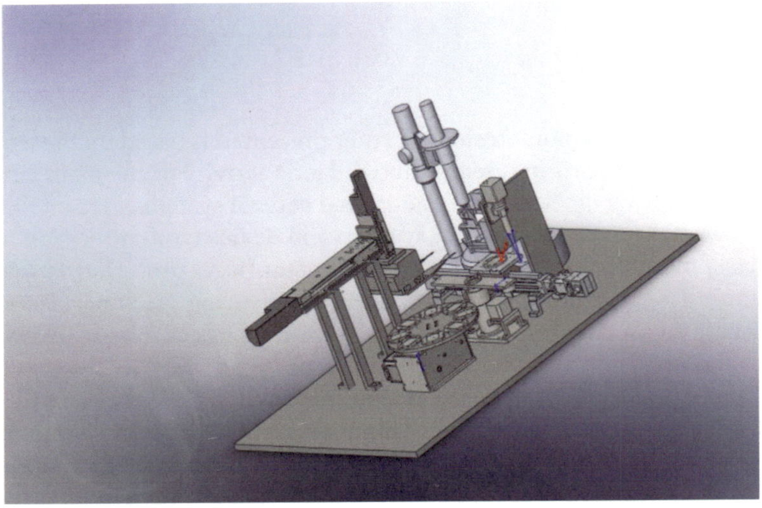

Figure 3.
Optoelectronic measuring and calibration system assisted by laser and AFM control.

The very thin nano-device calibration requires dedicated operational procedure for handling, and it is using support parts manipulated by a precision linear displacement system. These systems transfer the nano-devices by specialized automatic options (robot) to different precision measurement systems—optoelectronic, laser, or AFM—for calibration [3].

As shown in **Figure 3**, the equipment comprises a rotary feeding system, on top of which are placed eight support tables. On each support table, there is a specific plate support where nano-devices are introduced to follow the calibration procedure.

Experimental equipment includes mechanical, optical, and optoelectronic sub-ensembles, the optoelectronic measurement sub-ensembles, and the algorithms related to (real time) measurement system data acquisition, data processing, and measurement protocol presentation.

Control equipment (**Figure 4**) uses a rotary handling system including a feeding robot manipulator, a precision linear moving system, an optical measuring system, a laser measuring system, and a measuring system equipped with an atomic force microscope (AFM) [3].

The technical features of this experimental model ensure the following precision by:

- Displacement accuracy of the ultra-accurate-controlled linear positioning systems, 0.2 nm
- Laser measurement resolution, 1 nm
- AFM characterization resolution, less than 0.5 nm
- Optoelectronic measurement resolution, 10 nm

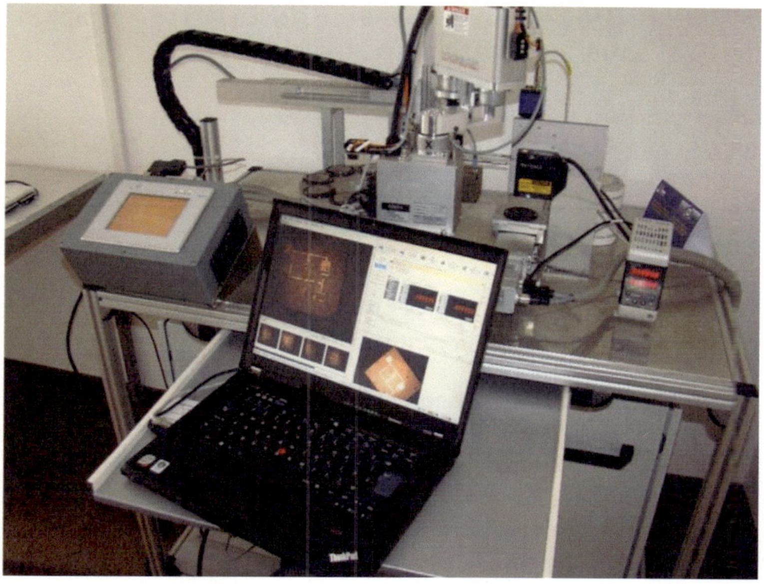

Figure 4.
The experimental equipment model.

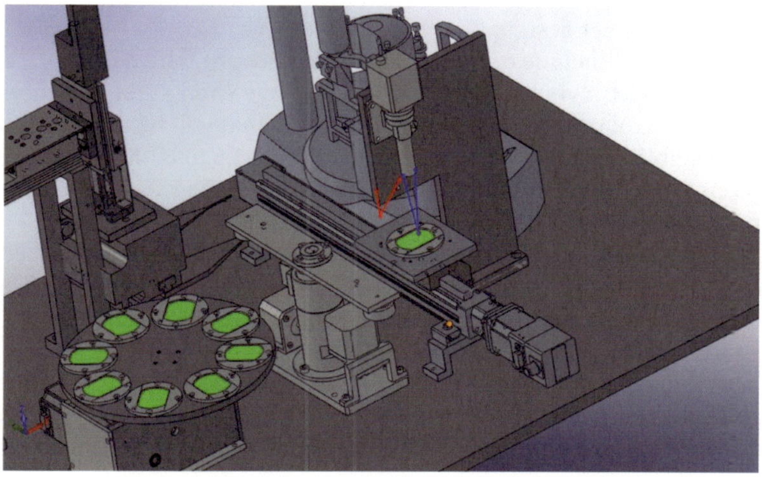

Figure 5.
Optoelectronic control.

Calibration procedure secures the nano-device optimal positioning in the support dedicated plate, which is precisely fixed on the support table from the rotary feeding system by computer coordination and transfer manipulator, which allows adjustments of the table during the precision displacement. For measuring operations with an AFM, a special feeding robot is used in order to keep accurate calibration characteristics. Dedicated computer programming procedure of the calibration system decides if the device is accepted (qualifying as good) continuing the production flow or is rejected (is not respecting the quality required) to scrap boxes (**Figure 5**).

The nanotechnological process can be adjusted using this equipment by programming automatic calibration for one, two, or three posts from eight available table supports [13]. The control software is versatile ordering automatic calibration process or manually controlled by a computer system or using touch screen applications.

The equipment developed in research institute INCDMTM is equipped to perform the flow control by means of three specialized systems:

- Optoelectronic control (microscope with CCD camera)
- Laser control
- AFM control

In this chapter the integrated control processes for nano-production flow using these three dedicated systems are presented summarily.

2.1 Optoelectronic control (microscope with CCD camera)

This testing and controlling method permits to check up the quality and cohesion of different nano-devices: semiconductor devices (SMD discrete components), microelectronic circuits, micromachined circuits, printed microcircuits, microsensors, and transducers. The optoelectronic control method using CCD camera may be adopted for finding defects from handling, assembling, or encapsulating all types of devices listed above [3] (**Figure 6**).

The equipment used in this control process must be able to demonstrate the quality conditions of the devices mentioned in accordance with the requirements envisaged in the product design.

Equipment should include optics (optical microscope) with a magnification range of 1.5–20 X with a view area accessible and large enough for determination of details. The control procedure sets up the devices that will be examined by producers at established magnifications ranging from 1.5 to 20 X. Measurements of dimensions (length/width/diameter) will be made using the order of the same range of magnification that provide good accuracy of measurements.

The dimensional measurements with optoelectronic microscope are ranked (width, length routes, or contacts) in the range of 10–2000 μm, and measurement resolution is 10 nm. The system offers rigorous linear movement of the sample nano-device based on two perpendicular directions at a distance of at least 5 mm (matching the test plan). In this case, measuring resolution should be less than 100 nm (**Figure 7**).

Applications that require the optical control are verification of integrated circuits, verification of printed microcircuits, and verification of microsensors based on amorphous magnetic materials.

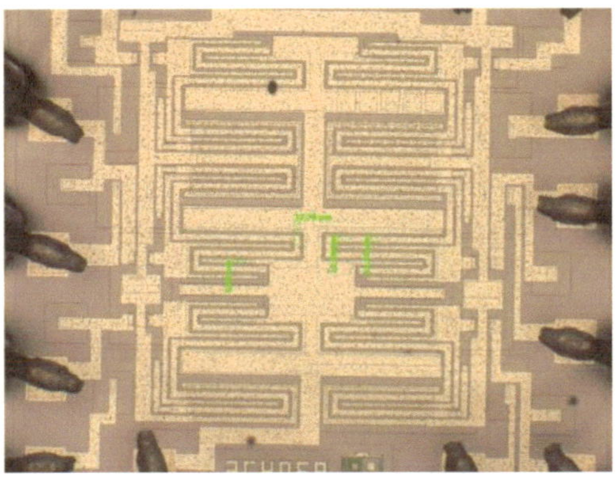

Figure 6.
Integrated circuit.

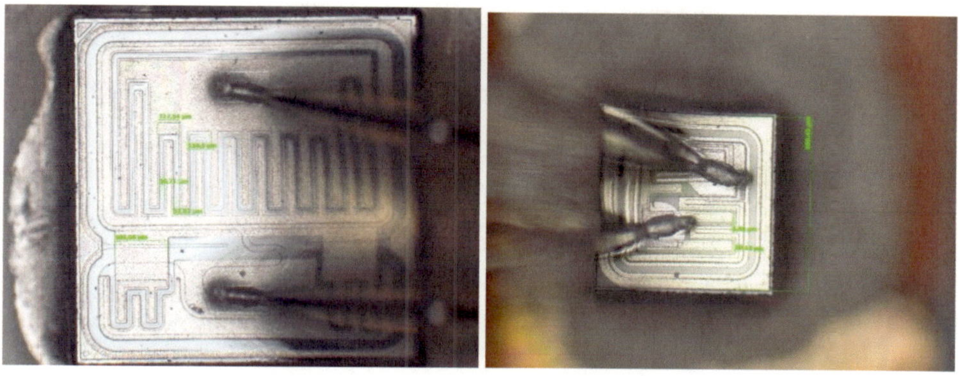

Figure 7.
Microprocessed circuits.

Some examples of the optoelectronic control applications in microelectronic circuits, micromachined circuits, printed microcircuits, and microsensors are shown (**Figures 8–13**).

Verification of integrated circuit (**Figures 8** and **9**) procedure includes:

- Verification of routes and establishing dimensional variations from the geometry of the proposed design by comparing with a theoretical form
- Verification of junctions and contacts
- Verification of profiles

Verification of printed microcircuit (**Figures 10** and **11**) procedure includes:

- Controlling the framing of deviations from the theoretical geometric shape of a circuit within the prescribed limit
- Control of circuit breaks
- Control of the geometric shape of the circuit
- Control of the presence and correct positioning of the components on the circuit
- Dimensional component control

Verification of microsensors based on amorphous magnetic material (**Figures 12** and **13**) procedure includes:

- Dimensional sensor control.
- Control of each sensor component.
- Control the correct positioning for each sensor on the circuit.
- Control the alignment of each sensor in the circuit.

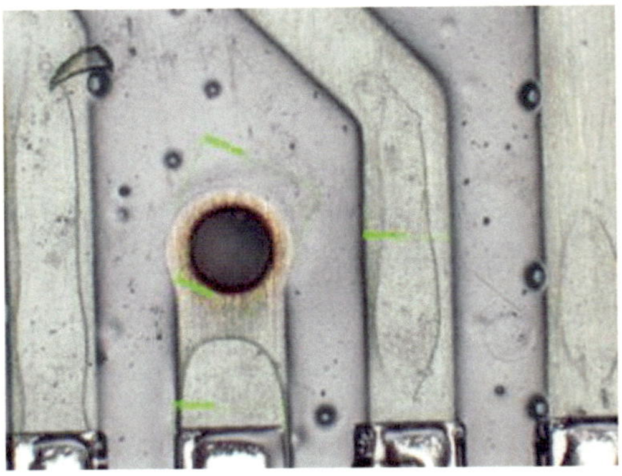

Figure 8.
Microelectronic circuits.

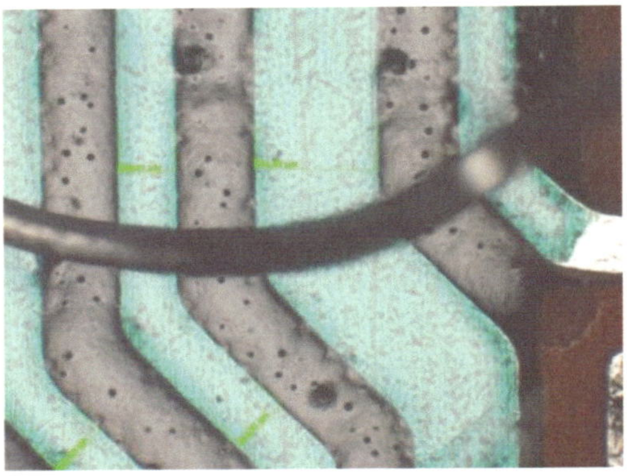

Figure 9.
Micro-imprinted circuits.

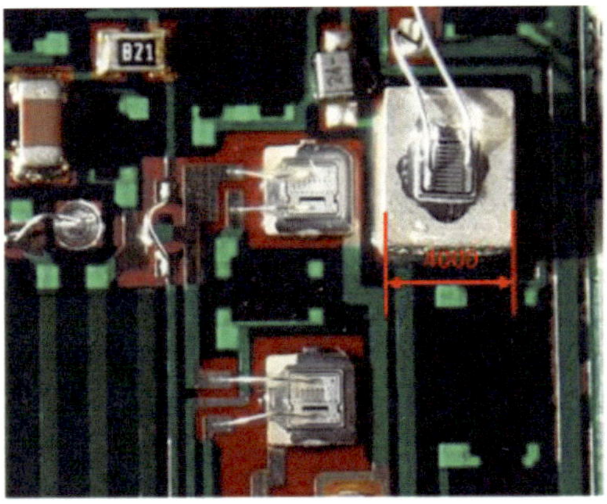

Figure 10.
Microcircuits.

2.2 Laser control

Laser measurement technologies gradually developed using multiple measurement principles that allow a large control flexibility and applicability for measurement and checking procedures.

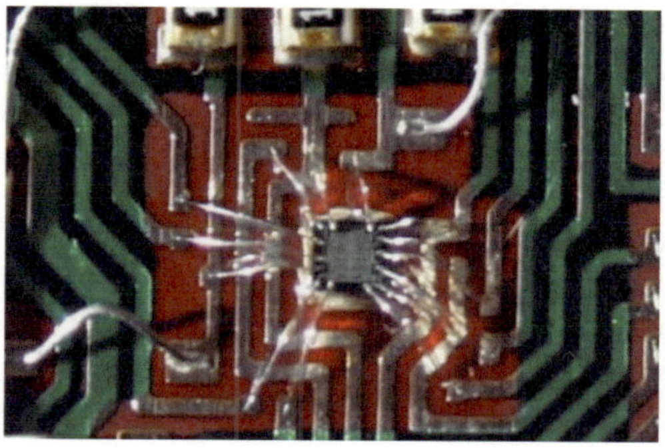

Figure 11.
Microcircuit.

Figure 12.
Cable routes and metal-plated holes.

Figure 13.
Inductive sensor.

Laser telemetry measurement principle offers a great variability of distance measuring systems up to kilometer lengths. The INCDMTM center developed applications to measure distance by telemetry satellites during the formation flying useful to maintain and adjust flying positioning. This application allows monitoring of distance length between satellites, and it controls trajectory of each satellite for keeping formation flying.

Interferometry principle is used for measurement covering distances of 80–120 meters and allows high resolution up to 0.01 nm. With a large experience in the development of checking and measurement applications for sensors, transducers, coordinate measuring machines, and precise CNC machines, we proposed to use the interferometry principle for nanotechnology processes where very precise displacements in a network system can be supervised with specific sensors. Laser triangulation is an accurate measurement principle with resolution of 1 nm. This method uses measurement referential to a point for distance and object presence determinations or referential to a line covering 3D forms and a profile dimension.

The new equipment designed for very precise measurement within time checking methods applies triangulation principle. Nevertheless our institute developed measuring systems using triangulation method 30 years ago [1]; in this case to assemble this measuring equipment on the nano-production flow, we acquired some measuring systems from a specialized company.

The purpose of using this method of control is to check the quality conditions of the semiconductor devices (discrete component-type SMD) of microsensors and transducers (e.g., control surfaces, movement control, control distance/size, position control, etc.).

The equipment used in this control must be able to demonstrate the quality conditions of the devices mentioned in accordance with the requirements envisaged in the design. It should include laser equipment and devices enabling precision movements on three axes (nano-positioning stage).

The measuring principle is the method of triangulation, having a measuring range of 5 mm, measurement resolution of 1 nm, and laser measurement resolution of 1 nm.

The value of the dimensions (width/length/height of routes, etc.) that can be checked is in the range 1–2000 μm. The system allows linear movement of the sample in three directions perpendicular to distances of at least 5 mm with nanometer precision. The precise positioning table is fixed in the laser calibration position as shown in **Figure 14**.

One of the applications appropriate for using laser-based measuring method is verification of integrated circuits, procedure that includes:

- Verification of routes and dimensional variations of the geometry identified to the proposed design by comparing with a theoretical form

- Verification of junctions and contacts

- Verification of profiles

Some examples of electronic microcircuits where laser control is applicable are shown in **Figures 15–17**. First, one control application defined by triangulation method for measuring and verification of the profile, positioning, and present splice of a pin in the integrated circuit is presented in **Figure 15**.

In microcircuit manufacturing, one important issue raised by specialists is the presence and correct checking position of each specific component to ensure the designed function of integrated circuits. This checking is presented in **Figure 16(a)** using a Keyence scanner. Continuous trends of minimizing the characteristic dimensions in integrated circuits and the rapid multiplication of functions determined for the same products lead to specialized very fine and narrow circuits' paths. Each circuit's path must to be produced respecting some rigorous requirements: minimum dimensions, distance between two of the closed paths, and the transversal profile of each path line. If the checking of circuit's form and the correspondence with theoretical design of the final product is realized with optoelectronic methods using video inspection, the verification of integrated circuit parameters is made by laser triangulation as shown in **Figure 16(b)**.

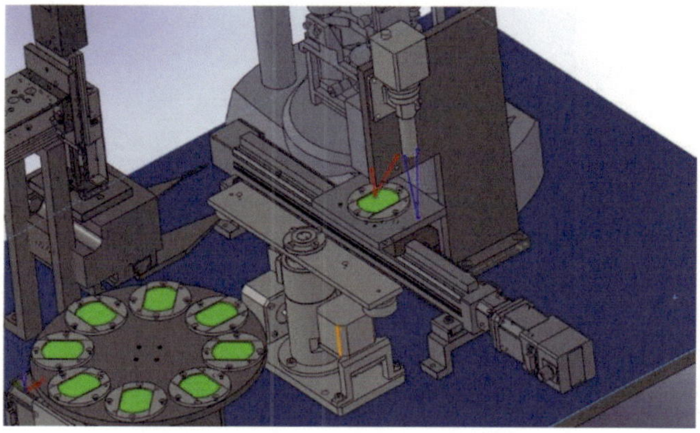

Figure 14.
Calibration with laser.

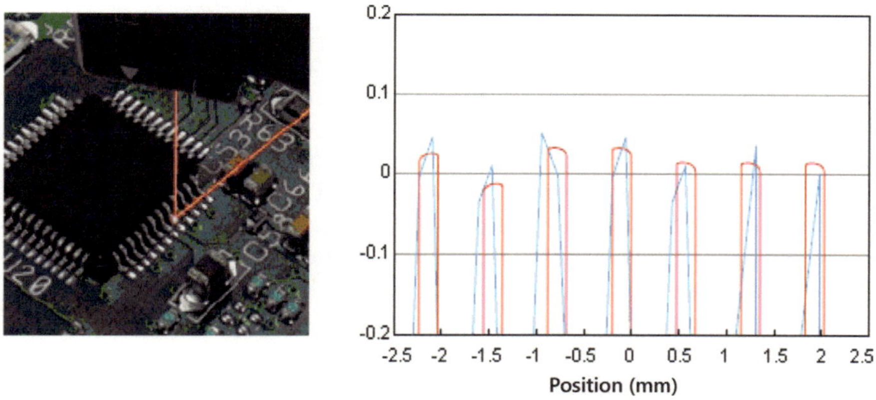

Figure 15.
Verification of the profile of pins for an integrated circuit.

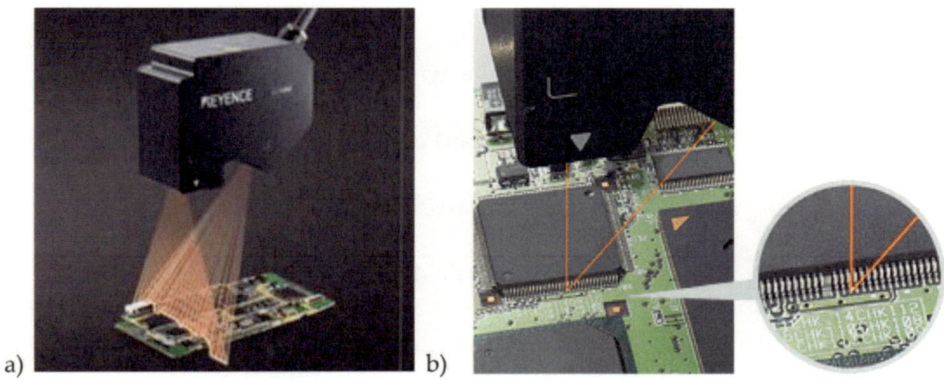

Figure 16.
Verification of the profile of microcircuits (a) and integrated circuits (b).

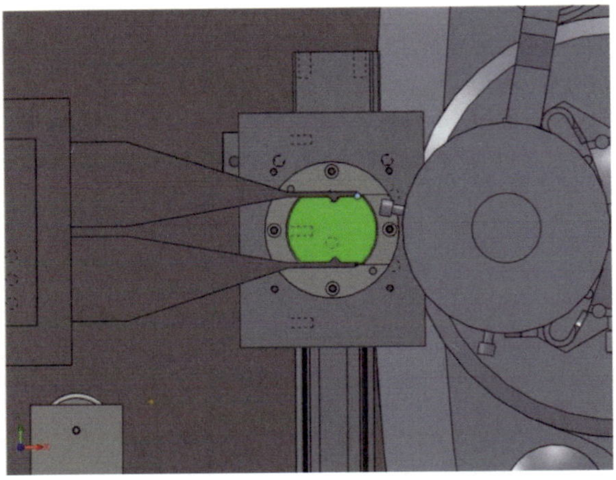

Figure 17.
Gripper clamps up the nano-device support.

2.3 AFM control

One of the problematic issues in the nano-production line control is the automatic maneuver of nano-devices. To protect nano-devices (microcircuits) during checking operations, there are specific item supports used with automatic precise displacement. Therefore, considering the nano-device control procedures, all parameters settled for measurement can be easily provided in each control point of the new equipment (optical, laser, and especially for atomic force microscope, AFM, characterization).

The purpose of using this AFM control method for nano-device testing and inspection plays an important role, and it is appropriate to check and to keep right conditions of integrity and quality of porous alumina membranes (alumina template) having pore sizes included in the nanometer range.

This control method can be used for inspecting defects that may result from the production (manufacturing industry), handling, or assembly of alumina membranes [9]. Control equipment used must be able to demonstrate the quality conditions of the porous alumina membranes in accordance with the requirements envisaged by product theoretical design. Control equipment is endowed with specialized systems that include an AFM.

The method of verification, control, and calibration includes the following procedure characteristics:

- The device must provide noncontact imaging solutions for nanoscale metrology.
- The scanning range on XY (sample plan) must be at least 100 × 100 μm.
- The scanning range on Z must be at least 25 μm.
- The values of dimensions (width/length, diameter pores, etc.) that can be checked are in the range of 1–500 nm.
- The system allows motorized sample stage in three directions, at a distance of at least 5 mm.
- The measuring resolution must be higher than 0.5 nm.

DOI: http://dx.doi.org/10.5772/intechopen.80425

Applicable for AFM control are porous alumina membranes used for obtaining nanowires via the electrode position process, and the procedure includes:

- Verification of the dimensional pore width and of the geometry of the proposed design, compared with theoretical form.
- Verification of membrane integrity.
- Verification of profiles.

Checking and control method using AFM is described summarily as follows: nano-devices are positioned on the support plate which is useful for automatic maneuver. Maneuver operations are driven by a robot that has a clapping system for interoperation displacements of support plates (**Figure 17**).

Before automatic operation offsets, it is required to set up the measuring head of AFM, which is equipped with a special support fixed on the adjustment unit of this head. The special support must be provided with guidance systems for support plate corresponding to each nano-device that must be calibrated. Adaptation of AFM measuring head position is settled following two rectangular directions using two fine-pitch screws that allow micron precise positioning of the measuring head referential to laser beam. Setting up of detection systems is performed by using control software program.

Automatic feeding of support plates with nano-devices is completed by a precise robot using a special gripper with fine claw clamps that hold and fix the support plates with nano-device in specific hole. Gripper form allows maneuver and fixing of nano-device support compatible with the guiding system of the special support from AFM measuring head.

One important issue regarding AFM operation is the laser beam alignment into the cantilever. For nano-device precise positioning for calibration procedure, a universal measuring head was selected. The laser beam alignment is realized by joist displacement in cantilever in relation to the beam spot.

The gripper clamps up the support table with the calibration nano-device, and it is built to introduce the nano-device fixed on its support plate ready to be verified directly in the AFM socket without protection cap removal (**Figure 18**). This automatic process using AFM control admits time-saving and more productivity.

The robot lifts the support with the nano-device that must be calibrated to the height of the positioning socket of the AFM and introduces that support in the right position (**Figures 18** and **19**).

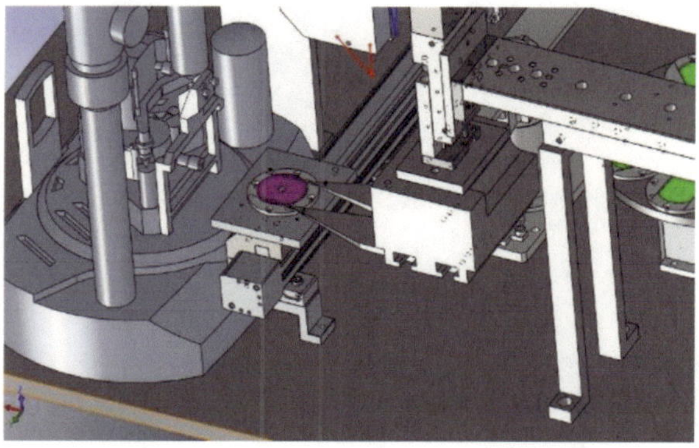

Figure 18.
Robot lifting nano-device related with AFM positioning socket.

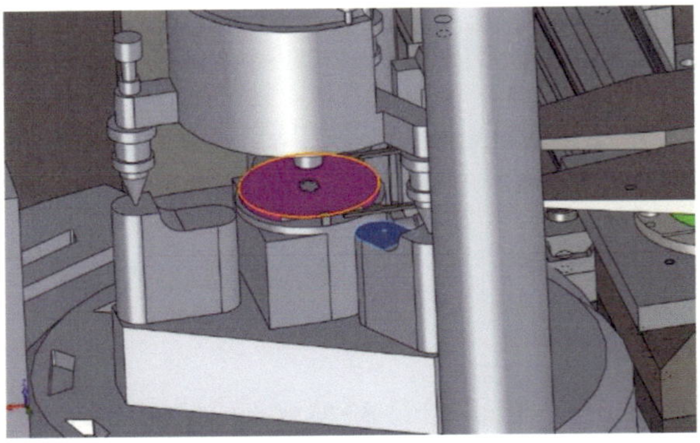

Figure 19.
Robot positioning nano-device support plate in the AFM socket.

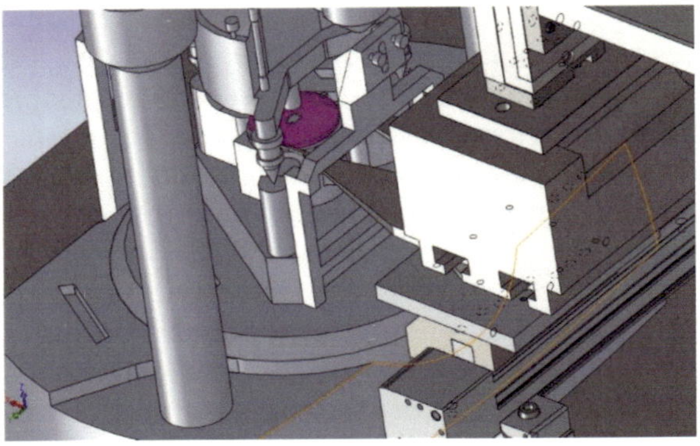

Figure 20.
Robotic performance of AFM calibration.

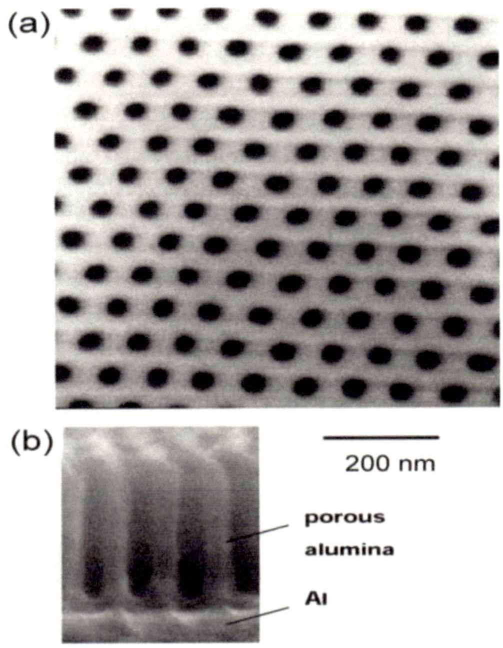

Figure 21.
Porous alumina membranes with pore diameters at nanoscale (SEM image) (a) top view and (b) lateral view.

The robot clamps off the gripper and takes down the support with the device that needs to be calibrated, and the AFM catches and holds the nano-device in the calibration position (**Figure 20**).

Control and checking technology defines if the verification is done in each checking position or only in a few checking positions and ensures the monitoring of settled characteristics important to be verified in every checking position. After every finalized checking operation, the equipment decides if the nano-device is within the tolerance limits and if it is accepted or not. If a specific characteristic of every nano-device is not corresponding within the settled limits of the technology that is designated by the measurement program, the feeding robot receives the command REBUT in every checking position, and the nano-device is eliminated from the production flow [13]. If the nano-device is accepted according to settled limits for each checking position, the robot receives the command GOOD, and the nano-device is introduced ensuing further into production flow.

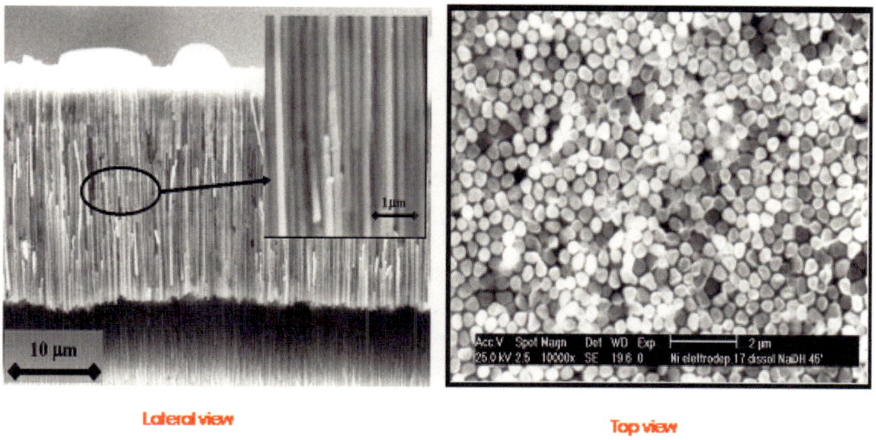

Figure 22.
Nanowires obtained through the process of nano-disposition in porous alumina membranes (SEM images).

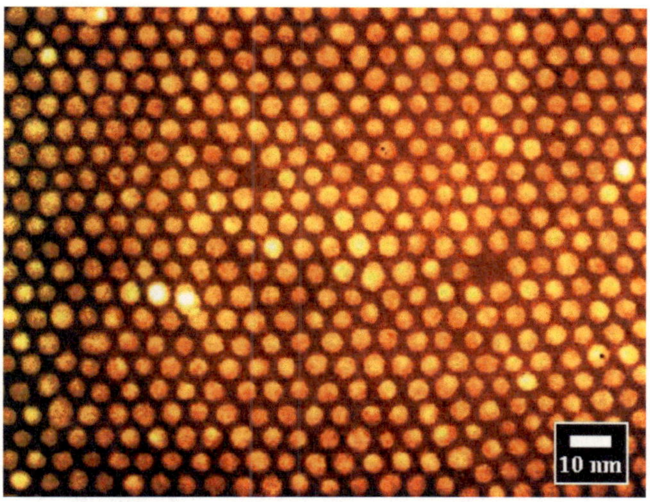

Figure 23.
Nanowires in the porous alumina membrane (AFM image).

Some relevant pictures present examples of AFM control applicability to porous alumina membranes where pore diameters are in the range of nanometers (**Figures 21–23**).

The images from **Figures 21–23** are achieved in the National Institute of Research and Development for Technical Physics, Iasi, Romania, during the collaborative research using scanning electron microscope (SEM) and AFM. These applications are demonstrative for AFM characterization and are dedicated for very precise control and checking processes in the nano-production flow.

3. Conclusions and future research

The experimental model permits optical, laser, and AFM microscopic verifications of realized nano-devices in order to correct possible production errors [3, 6, 12, 13]; thus, it allows nano-production calibration and automatic selection of rebuttal during flow processes for dedicated dimensional control in range from less than 1 nm up to micrometers or millimeters.

Future research will aim at the development of detailed technologies for various applications in nano-device production field that need to be calibrated covering all ranges of electronic nano-devices, optical nano-devices, biological nano-devices, nano-materials, and nano-sensors.

To evolve from laboratory stage to nanotechnology production lines, more research and innovations may allow over passing the actual barriers:

- Traditional measurement techniques used for normal dimensional characterization cannot be applied to nano-structures.

- Special rules and standards must be introduced for nano-structures and nano-material characterization reducing errors in inspection and quality checking procedures.

- More innovative equipments must be projected in order to solve the mentioned issues.

- Different specific studies for new equipments for control production regarding nano-structure proprieties should promote reproducible production of nano-structures and nano-materials.

Nanometrology opens opportunity creation of international standards and equipments for calibration of the products and equipments used in industrial production and offers more chances of new scientific discoveries regarding innovative commercial products.

The future development of nanotechnology cannot be achieved without progress in ensuring a well-controlled, stable production carrying dimensional control and in terms of other quality characteristics. This depends both on the strategy of each area of development and especially on the joint development of the nanotechnology field. First, research should be coordinated and developed in collaboration with companies, and secondly research for standardization in the field of nanometrology must be promoted by government programs. Efforts need to be united between those with common concerns for the progress of the nanotechnologies in precise industry.

For unitary development and interchangeable products, rules and standards need to be created at European and international status both for the acceptance of production and for systematization of products and components at the nanoscale. For large-scale production of nanotechnology products, equipments must be developed for both industrial production and production control. In order to ensure stable production, international rules must be developed and immerged for calibration of production flow and for calibration of control equipment production. Another issue that needs to be considered is that of environmental production conditions. We need to rethink environmental standards for this type of production. The old classification and standardization of clean rooms no longer correspond, and it is necessary to improve the clean room technical standards and add specific parameters. At the atomic force microscope (AFM), the measured parameter value is drastically influenced by its position relative to the air circulation system, noise, and vibration not only of temperature and humidity.

Author details

Gheorghe Popan* and Ana Elisabeta Oros Daraban
National Institute of Research and Development in Mechatronics and Measurement Technique (INCDMTM), Bucharest, Romania

*Address all correspondence to: popangeorge@yahoo.com

References

[1] Popan Gh, Sorea S, Savu T. Mechatronic System for Scavenge the Laser Ray. Romanian Review Precision Mechanics, Optics & Mechatronics. 2007;**31**. ISSN 1584-5928

[2] Chiriac H, Popan G, Ursu D, Gheorghe M, The Importance of the Production Calibration in the Nano-domain. Romanian Review Precision, Mechanics Optics Mechatronics. 2008;**33**. ISSN 1584-5928

[3] Popan G, Angelescu D. Flexible control system used in the nano-technological production flow. In: Proceedings of 5th International Conference on Advanced Manufacturing Engineering and Technologies (NEWTECH 2017), Lecture Notes in Mechanical Engineering. pp. 67-77. DOI: 10.1007/978-3-319-56430-2_6

[4] Leão C-P, Soares F-O, Machado J-M, Seabra E, Rodrigues H. Design and development of an industrial network laboratory. International Journal of Emerging Technologies in Learning. 2011;**6**(2):21-26. DOI: 10.3991/ijet.v6iS1.1615

[5] Chiriac H, Popan G, Ursu D, Gheorghe M. Nano-technological measuring systems. In: Proceedings of the Second International Conference on Innovations, Recent Trend and Challenges in Mechatronics, Mechanical Engineering, New High-Tech Products Development (MECAHITECH `10), Bucharest. 2010. pp. 486-491, https://incdmtm.ro/editura/imagini/Proceedings%20of%20MECAHITECH%202010.pdf

[6] Donmez A. EL Program: Smart Manufacturing Processes and Equipment, National Institute of Standards and Technology, US Department of Commerce. 2013; https://www.nist.gov/sites/default/files/documents/el/isd/sbm/SMPE-ProgramDescription_2013.pdf

[7] Machado J, Denis B, Lesage J-J. A generic approach to build plant models for DES verification purposes. In: Proceedings of the Eighth International Workshop on Discrete Event Systems, (WODES` 06); 10-12 July 2006; Michigan. pp. 407-412

[8] Sorin S, Popan G, Atanasescu A. Nanotechnology and solar cells. In: Proceedings of the Second International Conference on Innovations, Recent Trend and Challenges in Mechatronics, Mechanical Engineering, New High-Tech Products Development (MECAHITECH `10), Bucharest. 2010. pp. 473-476

[9] Barros C, Leão C-P, Soare F, Minas G, Machado J-M. RePhyS: A multidisciplinary experience in remote physiological systems laboratory. International Journal of Online Engineering. 2013;**9**(Sppl. 5):21-24. http://online-journals.org/index.php/i-joe/article/view/2756/2650

[10] Popan G, Gheorghe G, Vieru A. Control procedures in the production flow. In: Proceeding of International Conference on Innovations, Recent Trend and Challenges in Mechatronics, Mechanical Engineering, New High-Tech Products Development (MECAHITECH `09) Bucharest; 8-9 October 2009

[11] Silva M, Pereira F, Soares F, Leão C-P, Machado J-M, Carvalho V. An overview of industrial communication networks. In: Springer Cham editor, New Trends in Mechanism and Machine Science. Vol. 24; 2015. pp. 933-940. DOI: 10.1007/978-3-319-09411-3_97

[12] Popan G, Palade D-D, Lung I, Tacutu I. Influence of constructive-functional elements of the laser head

systems accuracy. In: Proceedings of the 6th International Working Conference – "Total quality management- advanced and intelligent approaches", 7-11 June, Belgrade, Serbia Rev: 8843 Total Quality Management & Business

[13] Vieira G-G, Varela ML-R, Putnik G-D, Machado J-M, Trojanowska J. Integrated platform for real-time control and production and productivity monitoring and analysis. Romanian Review Precision Mechanics, Optics and Mechatronics Review. 2016;**50**:119-127. https://search.proquest.com/openview/390da36d8112b1fe4783f794cf3006aa/1?pq-origsite=gscholar&cbl=2035039

Chapter 3

The Design and Development of Control System for High Vacuum Deoxygenated and Water-Removal Glove Box with Cycling Cleaning and Regeneration

Ming-Sen Hu

Abstract

This study proposed a high vacuum deoxygenated and water removal glove box control system. Through parameter setting, the system can automatically perform various glove box cleaning operations and quickly reach the micro-oxygen and micro-water concentration requirements. In addition, two sets of reaction tanks are built in the system, and the hardware pipeline switching design and monitoring software control are used to provide two sets of reaction tanks to execute the cycling cleaning and cycling regeneration operation procedures synchronously, which can effectively solve the problem of interruption of the experimental process, improve the efficiency of its cleaning operations, and greatly reduce the manpower and material costs of the glove box operation. In addition, the system can automatically record the relevant data during various operations for the analysis of glove box monitoring effectiveness.

Keywords: glove box, control system, cycling cleaning, cycling regeneration, monitoring software, system integration, performance analysis

1. Introduction

A glove box, also known as an anaerobic station, is a vacuum environment without any water (H_2O), oxygen (O_2), and organic gas, where high purity inert gas is filled, and the active substances are filtered out, which can prevent the external personnel from direct contact, so that the materials placed in the glove box can be stored or tested in the vacuum environment free of water, gas, and oxygen, and the material is operated by people in a safe and contactless manner [1]. The glove box is used in a wide range of applications such as the scientific research of chemical/chemical engineering/material/drug, and so on; the research and manufacturing of organic optoelectronic OLED/polymer PLED light emitting displays; the manufacturing and research of lithium ion batteries/lithium polymer batteries/solar cells/high capacity capacitors; the research and manufacturing of special (HID) light bulbs; metal welding or laser metal welding process; the preservation or research and manufacturing of noble sensitive drugs and nuclear energy research; and so on [2–4].

The glove box is mainly composed of a glove box body, a vacuum system, a gas circulation exchange system, and a control system [5]. The glove box body is provided with an antechamber (also called a transfer box) and an isolation glove. Usually a viewing window is arranged on the front of the body, so that the operator can clearly observe the operating conditions inside the box, and the operation process can be intuitively displayed in front of the operator. The glove box facilitates operators with isolation gloves in an anhydrous, oxygen free, and vacuum environment. Basically, the vacuum glove box is an important application of vacuum technology [6–9]. What degree of vacuum the glove box can reach not only depends on quite different manufacturing costs, but also affects the effectiveness of vacuum preservation of sensitive items, and vacuum level is also an important key to effectively remove moisture and oxygen in the box. Usually, a high efficiency glove box must be the high vacuum box (transfer box) [10–12].

For high vacuum deoxygenation and water removal glove box, the oxygen content and water content in the box must usually reach the ppm level (i.e., up to 10^{-4}%) of the micro-oxygen and micro-water concentration [13–15]. However, it is difficult to achieve effectively such micro-oxygen and micro-water concentration. Usually, the glove box must be vacuumed, and then the box is filled with an inert gas (such as nitrogen or helium), and this type of evacuation and nitrogen (or helium) filling operation procedure must be performed multiple times. Thus, the deconcentration of water and oxygen can be accelerated, and the concentration of water and oxygen in the ppm range can be achieved more efficiently. In addition, the micro-water analyzer and micro-oxygen analyzer that detect ppm levels of water and oxygen concentration also have a detection range of ppm (e.g., 0–1000 ppm). To ensure the normal use of such analyzers, the input of water and oxygen concentration values of this type of analyzer should also fall within its detection range to avoid damage to the analyzer. If water and oxygen of high concentration (e.g., percentage level) are input for a long time, these analyzers will be easily damaged and cannot be used any more.

On the other hand, the transfer box of the glove box system is mainly for the user to put the materials into the glove box body or take out the materials from the box body. However, when the glove box body has reached the micro-oxygen and micro-water concentration, the material should be placed through the transfer box. The material must be first put into the transfer box, and then the transfer box is sealed and evacuated to make the water and oxygen content similar to the glove box. But the transfer box in the vacuum state will suck, making it difficult to operate, so the inert gas in the glove box must be introduced into the transfer box so that the pressure is balanced, and then the isolation glove can be used to place the material into the box. On the contrary, if the material is to be taken out of the glove box and placed back in the transfer box (to be removed from the transfer box), the transfer box must be evacuated first, and then the inert gas must be introduced from the glove box to balance the pressure.

Although the glove boxes commonly used in various industries currently have a control system, they mainly monitor the vacuum system and the gas circulation exchange system through PLC in a semi-automatic manner [15, 16], and they must be operated by people by starting and setting the parameter of the vacuum system or gas cycling exchange system, so that the vacuum operation or gas cycle cleaning operations of such a glove box often need to wait for the completion of the parameter setting for the next stage, which tends to cause inconvenience in use and operation; on the other hand, to analyze the effectiveness of the vacuum operation and cleaning operation of the glove box system, it is necessary to automatically record the required measurement information for processing during each operation. However, no major glove box system is currently available with automatic recording of measurement data. The PLC control system is also very inconvenient for the instantaneous recording and processing of measurement data. Therefore, the existing glove box systems do not provide users with efficiency analysis function for vacuum operation and cleaning operation.

Furthermore, since the commonly used glove boxes are equipped with only one gas circulation exchange system, namely, a reaction tank capable of removing moisture and oxygen through the reagent [15, 16], when the chemical reagent in the reaction tank is used up, the operation must be stopped. After the cleaning reagent is replenished, replaced, or regenerated, the cleaning operation can then resume. The regeneration of the reaction tank is usually by burning a low concentration of hydrogen to restore the cleaning function of the reagent [16]. However, during the experiment or operation in the glove box, due to the slight leakage of the glove box or some gas generated by the experimental operation, the internal environment of the box will constantly change, so that the box must be continuously monitored and the cleaning operation must be performed when the concentration of water or oxygen exceeds the specified value [11, 17]. Therefore, when the glove box system is only configured with one reaction tank, except for the inconvenience of use and operation, it may easily lead to the problem of interruption of the experiment or the operation process, thereby affecting the quality and efficiency of the glove box. Although there are currently a small number of larger glove box systems that can be configured with two or more sets of reaction tanks [16], the configured redundant reaction tanks are usually only alternative replacement devices. When the reagent of the reaction tank is used up and the cleaning function is lost, it must be switched to other reaction tank to continue the cleaning work by manual operations. In addition, the operation of replacing or regenerating the reagent is performed on the reaction tanks that have lost the cleaning function. However, this operation method still cannot solve the problem of interruption in the experiment or operation process.

In view of the defect of experimental or operational interruptions in the use of the glove box, a cycling cleaning regeneration mechanism was proposed in this study [18, 19]. This mechanism is mainly to build two reaction tanks A and B. At the beginning, the reaction tank A performs the cleaning work, and the reaction tank B takes the rest. When the reaction tank A loses the cleaning function due to the use up of the reagent, it will automatically switch to the tank B to continue to work, and the regeneration operation of the tank A starts at the same time. When the tank B loses the cleaning function due to the use up of the reagent, it immediately switches to the operation of tank A and simultaneously performs the regeneration of tank B. Thus, the two reaction tanks are used alternately to perform the cycle cleaning and recycling operation, which can effectively solve the problem of interruption in the experiment or the operation process of the glove box system.

Based on this cycle cleaning regeneration mechanism, this study developed an advanced control system for high vacuum deoxygenation and water removal glove box. Through the design of switching by the hardware pipeline and control by the monitoring software, the system provides transfer box cleaning, glove box cleaning, vacuum preservation, end of preservation, cycling cleaning regeneration, and other basic operating functions on the one hand, and each operating function can be automatically executed through the setting of the control parameters and the instrument parameters; on the other hand, during the execution of each operating function, the system can automatically record all the measured data and elastically display the vacuum test curves and the cleaning test curves. This can provide the user with the analysis of the efficiency of cleaning operation and vacuum operation of the glove box. The results of this kind of performance analysis can be used as a quality parameter to determine the quality of the glove box and the basis for the user of the glove box to plan experiments or operations.

To verify the function of the deoxygenation and dehydration of the glove box system developed, a hydrogen storage and cleaning mechanism was built in the glove box of this study to serve as a specific application of the glove box system, and the application functions of hydrogen storage and hydrogen cell cleaning were provided in the monitoring software. This kind of design mainly considers that hydrogen is the main fuel for hydrogen vehicles in industry, and usually needs to be stored in a hydrogen storage cell. The hydrogen storage operation of the hydrogen storage cell must be in an oxygen free environment, including prior vacuumizing to remove the air in the hydrogen storage cell, then filling the hydrogen storage cell with hydrogen in an oxygen free environment, so as to avoid the danger of explosion due to the combination of hydrogen and oxygen. Therefore, we use this micro-oxygen and micro-water glove box as the operating environment of the hydrogen storage cell, providing operators with the ability to safely and smoothly perform the operation of evacuating the hydrogen storage cell and storing hydrogen for hydrogen storage purpose.

2. System design

The system architecture of the high vacuum deoxygenation and water removal glove box control system developed in this study is shown in **Figure 1**. In the figure, the solid blue line represents the gas flow path, the green dotted line represents the

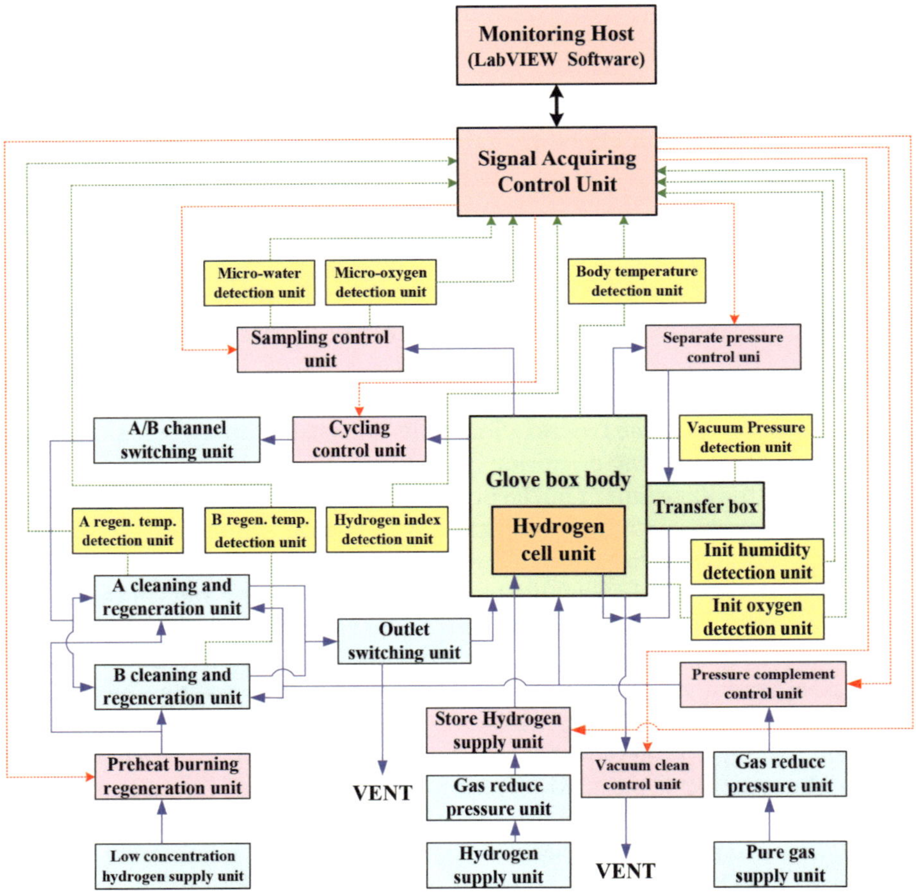

Figure 1.
Architecture of the glove box control system.

detection signal transmission path, and the red dotted line represents the control signal line.

The functions of each component unit in **Figure 1** are described as follows:

1. Glove box body: a box body of oxygen free and water free environment that provides a high vacuum for the user to perform material handling or storage. The body contains a set of hydrogen cell units that can store hydrogen or remove hydrogen.

2. Transfer box: it is used for users to put the material into the glove box or remove the material from the box.

3. Body temperature detection unit: the temperature sensor used to detect the temperature of the glove box body, and the detected temperature can be transmitted to the monitoring host.

4. Primary humidity detection unit: it can detect the humidity of the body in the atmospheric state, and it is a humidity sensor of the percentage range detection.

5. Primary oxygen detection unit: it can detect the oxygen concentration of the body in the atmospheric state, and it is an oxygen sensor of the percentage range detection.

6. Partial pressure control unit: it is a control module used to introduce the inert gas from the body into the transfer box to achieve a pressure balance.

7. Vacuum pressure detection unit: the pressure sensor used to detect the gas pressure (i.e., vacuum level) in the body, and the detected pressure can be transmitted to the monitoring host.

8. Pure gas unit: source of purified inert gas supply for high pressure nitrogen or helium.

9. Gas decompression unit: since the high pressure inert gas or high pressure hydrogen provided by the pure gas unit and the hydrogen supply unit are of very high pressure and cannot be directly used, the decompression unit must be used to decompress the high pressure gas before the gas enters the glove box system.

10. Gas supplement control unit: the control module to receive the command from the monitoring host, which is used to supplement the decompressed pure gas to the body or supply it to the regeneration reaction tank.

11. Vacuum removal control unit: the vacuum pump controlled by the monitoring host to remove the gas in the body or the transfer box and bring it to a vacuum state.

12. Micro-oxygen detection unit: a micro-oxygen detector that can detect ppm oxygen concentration. The detected oxygen concentration can be transmitted to the monitoring host.

13. Micro-water detection unit: a micro-water detector that can detect ppm level moisture concentration, and the detected water concentration can be transmitted to the monitoring host.

14. Sampling control unit: the sampling box controlled by the monitoring host may be used only for capturing gas in the body during the cycling cleaning operation for the micro-oxygen detector and the micro-water detector to detect the water and oxygen concentration so as to prevent such an expensive detector from being damaged by excessive concentrations of moisture or oxygen.

15. Cleaning regeneration unit: it consists of two reaction tanks that can carry out cleaning and regeneration functions. The cleaning function is to use the cleaning reagent to adsorb oxygen and water vapor that flows through the reaction tank. When the regeneration function is performed, low concentration hydrogen gas is introduced to regenerate its cleaning agent.

16. Cycle control unit: it can be controlled to perform the cycling cleaning and the cycling regeneration operations alternately of the two reaction tanks A and B.

17. A/B channel switching unit: it is used to switch the pipeline channel of reaction tanks A and B.

18. Preheated burning regeneration control unit: it can receive the command from the monitoring host for heating of the reaction tanks A or B and to remove oxygen and water vapor adsorbed on the chemicals by burning low concentration hydrogen to achieve regeneration.

19. Regeneration temperature detection unit: a temperature sensor that can be used to detect the heating temperature when the reaction tank performs a regeneration function. Each of the reaction tanks A and B has a temperature sensor. The detected heating temperature can be transmitted to the monitoring host.

20. Low concentration hydrogen supply unit: it provides 3% concentration of hydrogen and nitrogen mixed gas, and this low concentration of hydrogen can be burnt to activate the regeneration function of the cleaning reagent (catalyst) in the reaction tank.

21. Outlet switching unit: it is used to switch the flow path of the gas from the outlet of the reaction tank A or reaction tank B. When the reaction tank performs the cleaning operation, the outlet gas flows back to the glove box body, and the outlet gas is directly discharged when the regeneration operation is performed.

22. Hydrogen supply unit: the hydrogen source module for hydrogen supply.

23. Hydrogen storage control unit: the module unit is controlled by the monitoring host to store and release hydrogen into the hydrogen storage cell.

24. Hydrogen index detection unit: it is used to detect the hydrogen index in the box body and send it to the monitoring host. If the hydrogen index is too high, there is a danger of explosion. The monitoring host will stop the hydrogen storage operation.

25. Signal acquiring control unit: the Advantech USB-4711A multi-function data acquisition card [20] is used as a signal capture unit to convert the analog and digital signals captured by each sensor into digital data, which are then transmitted to the monitoring host for processing. In addition, the Advantech USB-4750 digital output control card [21] is used as a signal control unit to transmit the control signals generated by the monitoring host to each control unit to control each valve switch.

26. Monitoring host: it is a personal computer that can execute LabVIEW monitoring software and configures a touch screen as the user interface for the user to operate the system. The LabVIEW monitoring software provides users with various manipulation functions, and at the same time, it can automatically record the relevant data of various operations, for drawing and flexible display of various test curves.

In the hardware design part, the P&ID (Process and Instrument Diagram) of the glove box control system is shown in **Figure 2**, where R-A and R-B represent the reaction tanks A and B, respectively; PI is the pressure indicator; PT, TT, HT,

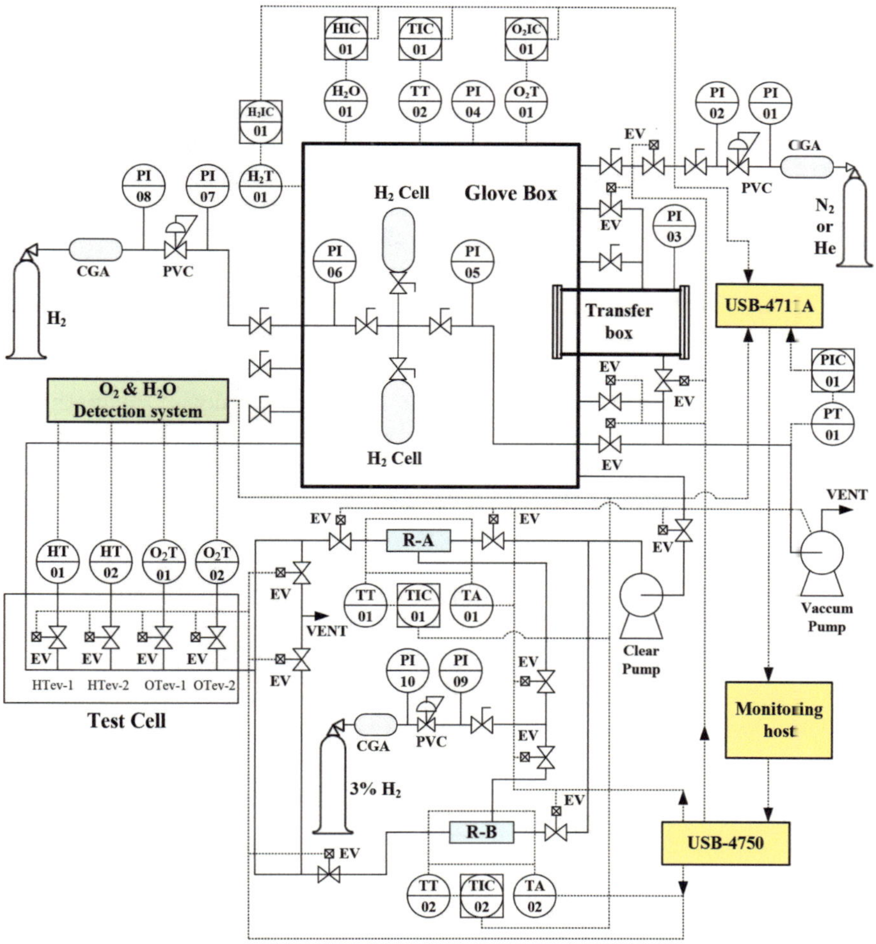

Figure 2.
Hardware design P&ID diagram.

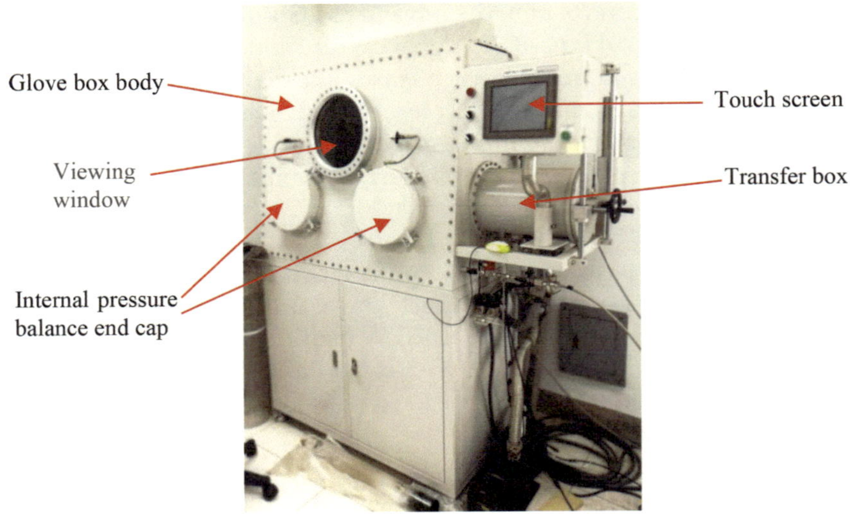

Figure 3.
Glove box control system entity.

O_2T, and H_2T indicate the detecting transducers of pressure, temperature, humidity, oxygen concentration, and hydrogen concentration; PIC, TIC, HIC, and H_2IC, respectively, indicate the indication controllers of pressure, temperature, humidity, and hydrogen concentration. All indication controllers inside the box can send signals to the USB-4711A adapter card and be processed by the monitoring host. TA and HA indicate the temperature and humidity actuators, respectively, and PVC indicates the pressure regulating valve. Each EV is an electric valve that can be controlled by the monitoring host via a USB-4750 adapter card.

To obtain high measurement quality of signals, such as temperature, pressure, water/oxygen concentration, and hydrogen concentration, etc., in terms of hardware, this study removed excessively large external signals through the masking technique and filtered the high frequency noises by virtue of short circuit filter capacitance. While in terms of software, this study employed the method of mean value to remove the influence of a small number of surges. For example, in a sampling period, the system captured 10 signals, calculated their mean, and used it as the measurement value, so that accurate and precise measurement signals can be obtained.

The glove box control system entity developed in this study is shown in **Figure 3**. We installed an internal pressure balanced end cap on the outside of the isolation gloves of the glove box to seal the isolation gloves inside the end cap. During vacuum operation of the box body, vacuum is applied together with the inside of the end cap to maintain the pressure balance. This avoids the pressure difference between the inside and the outside of the body, which may cause the isolation glove to inflate or even burst.

3. Software development

In this study, the automatic monitoring software for this glove box control system was developed using the LabVIEW graphical language [22–24]. The module hierarchy is shown in **Figure 4**. The monitoring software includes three major functional modules of parameter settings, program monitoring, and curve drawing. The parameter setting module can provide control parameter setting and instrument parameter setting functions. The program monitoring module can provide six

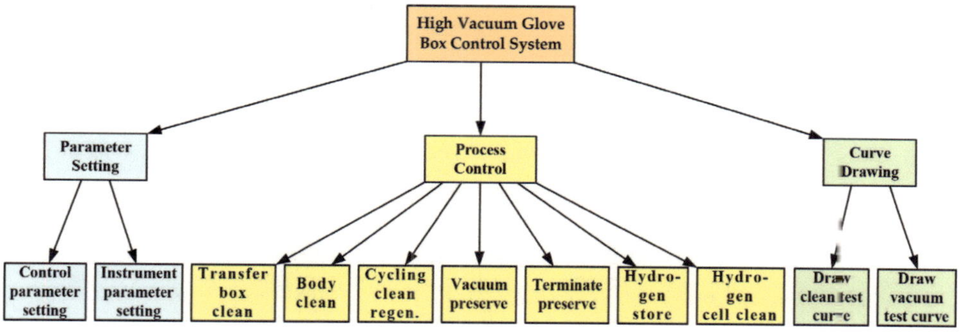

Figure 4.
The hierarchy of automatic monitoring software modules.

functions of monitoring, including transfer box cleaning, box body cleaning, cycling cleaning and regeneration, vacuum preservation, hydrogen storage tank cleaning, and hydrogen storage. And the curve drawing module can perform the function of drawing and displaying the cleaning test curve and the vacuum test curve.

Figure 5 shows the initial screen of the glove box automatic monitoring system designed by LabVIEW. Click the "Parameter Setting" button on the upper left of the screen to set the control parameters and instrument parameters. The seven green buttons on the right side of the screen are for the user to perform seven monitoring functions such as transfer box cleaning, box body cleaning, cycling cleaning and regeneration, vacuum preservation, end preservation, hydrogen storage tank cleaning, and hydrogen storage. Click the "Curve Drawing" button at the bottom left to draw and display the clear test curve and vacuum test curve.

3.1 Control parameter and instrument parameter setting

Before executing various control functions, the user can first set various control parameters and instrument parameters. The setting screen of control parameters is shown in **Figure 6(a)**. There are 16 control parameters that can be set in this screen. The control parameters set by users include the vacuum pressures of box body, transfer box and hydrogen cell, the set temperature of reaction tanks for cycling regeneration, the concentrations of water and oxygen set to stop cycling cleaning, and the times T1–T9 set to control various operation functions.

The instrument parameter setting can mainly provide the system designers to use the sensing instrument based on actual conditions to establish the software and hardware interface of the system, that is, the detection range of the sensor and the signal level relationship of the corresponding adapter card pin. The instrument parameter setting screen is as shown in **Figure 6(b)**. In the setting of instrument parameters, nine sensing ranges (for the regenerative temperature of reaction tanks A and B, the internal temperature of box body, the primary humidity and primary oxygen of box body, the concentrations of micro-water and micro-water, the vacuum pressure of vacuum pump, and the hydrogen explosion index) are given, and the corresponding signal levels of analog input pin AI0–AI8 are selected.

3.2 Real-time control process

After completing the control parameter setting, the user can perform various operation functions such as transfer box cleaning, box body cleaning, cycling cleaning regeneration, vacuum preservation, end of preservation, storing hydrogen and hydrogen cell cleaning, etc. The state transition diagram as shown in **Figure 7**

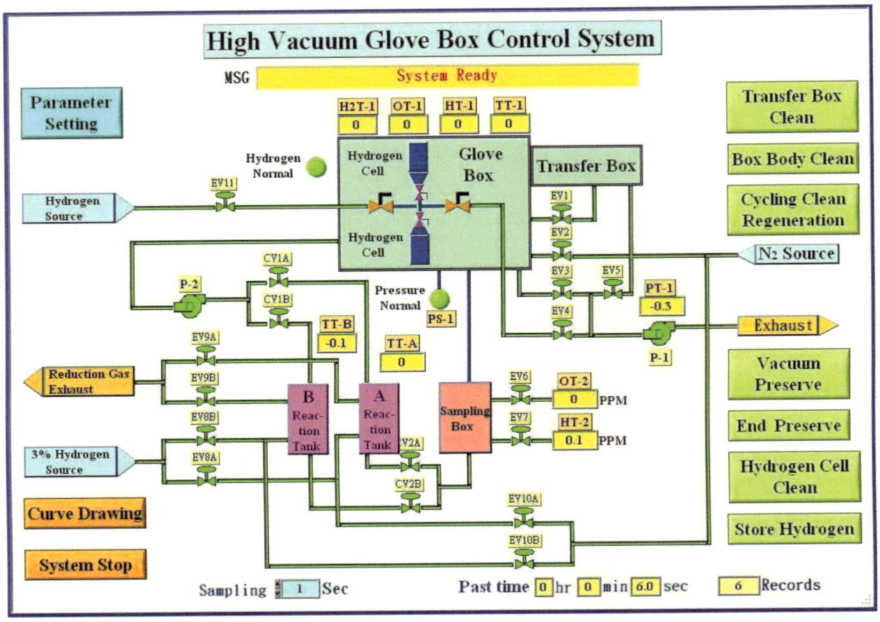

Figure 5.
Initial screen of the glove box automatic monitoring system.

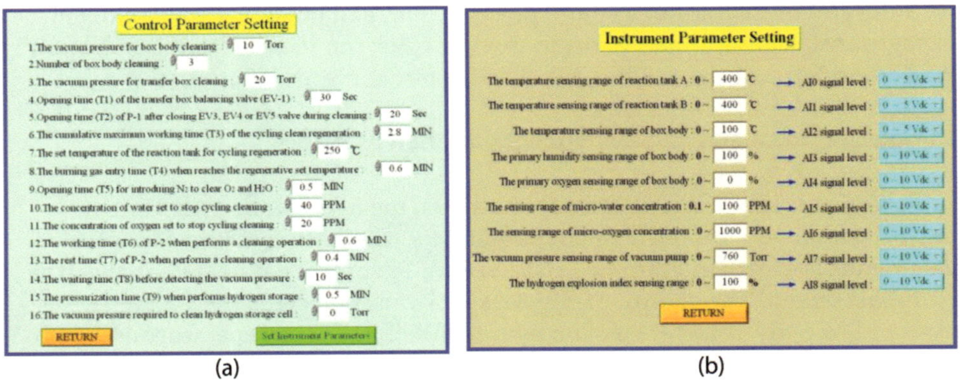

Figure 6.
Parameter setting screens. (a) Control parameter setting and (b) instrument parameter setting.

was used to represent the real-time control process for various operational functions of the glove box. The ellipses in the figure represent the states of the system, and the thick ellipses represent the final states. The double-circle ellipse represents the composite state, which is used to represent another state transition diagram. The line connecting the ellipses represents the transition of the state, and an event is attached next to each of the state transition connection line. This means that the transition of the system state is due to the occurrence of this event. In addition, a horizontal line can be added below the event. The action below the horizontal line is the action that accompanies the event, as shown below.

In the state transition diagram of **Figure 7**, when the system starts up, it enters a "Wait" state and enters the initial screen as shown in **Figure 5**, and the user can click "Transfer box clean", "Box body clean", "Cycling clean regen", "Vacuum preserve", "End preserve", "Hydrogen cell clean" or "Store Hydrogen" buttons to start the desired operation function. The system will enter the state corresponding

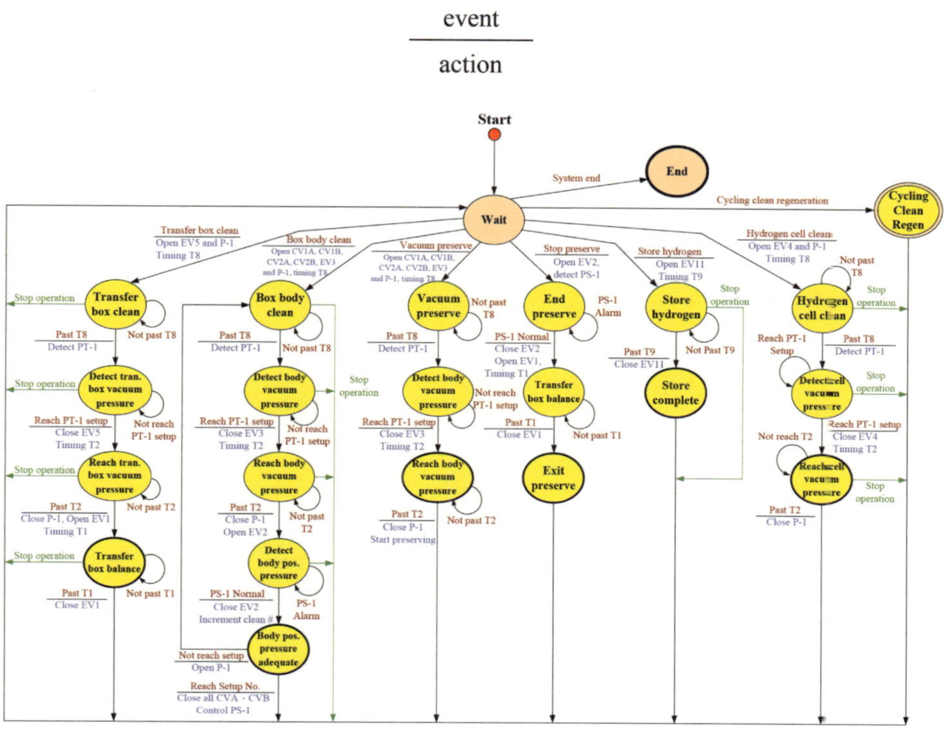

Figure 7.
State transition diagram of the glove box real-time control process.

to this operation function and start to execute its state transition diagram. The operating function of "Cycling clean regen" corresponds to a double-loop composite state. Entering this state will transfer to perform the "Cycling clean regen" as shown in **Figure 12**. After each state change graph of each operation function in the state transition diagram of **Figure 7** is executed, it will return to the "Wait" state for the user to select the next operation function, or the user may click the "System end" button to end the execution of this system.

3.3 Transfer box cleaning and box body cleaning operations

The transfer box cleaning operation removes moisture and oxygen from the transfer box of the glove box to a set vacuum level. When this operation is performed, the system first turns on the transfer box vacuum valve EV-5 and starts the vacuum pump P-1 to extract the air, and then enters the "*Transport box clean*" state, as shown in **Figure 8(a)**, which is the screen to execute the transfer box cleaning. In this screen, the red path represents the opening of the electric valve. Then it waits for T8 time, enters the state of "*Detect transfer box vacuum pressure*", starts measuring the vacuum pressure PT-1 (representing the transfer box pressure), and waits for the pressure to reach the set transfer box vacuum level (determined by the control parameters set in **Figure 6(a)**). When the vacuum pressure of the PT-1 is reached, it enters the "*Reach transfer box vacuum pressure*" state. At this time, the EV-5 shutoff action is started first, the T2 time is waited for, then the P-1 is closed, and then the "*Transfer box balance*" state is entered, this will open the transfer box balance valve EV-1 and time T1 will be counted, in order to introduce the inert gas in the box body into the transfer box, so that it can achieve pressure balance, thus completing the transfer box cleaning work.

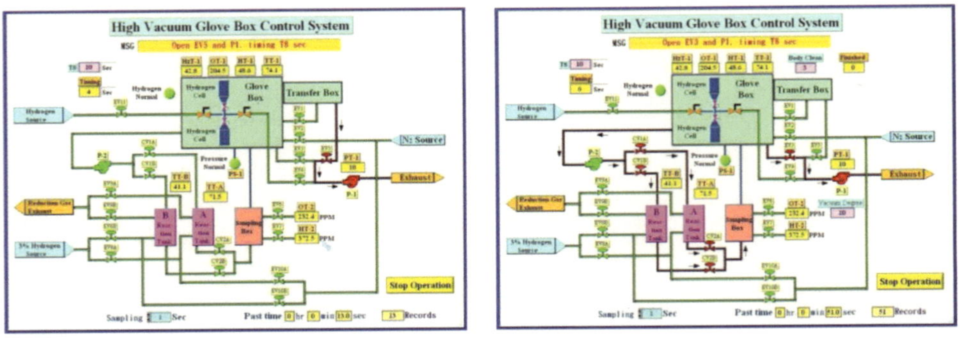

(a) (b)

Figure 8.
Execution screen of transfer box cleaning and box body cleaning. (a) Transfer box cleaning procedure and (b) body cleaning procedure.

The box body cleaning operation is used to remove the moisture and oxygen of the glove box itself so that it reaches the set vacuum level. When this operation is performed, the system first opens the box vacuum valve EV-3 and activates the vacuum pump P-1, and simultaneously opens the inlet valves CV1A, CV2A, CV1B, and CV2B of the regeneration system, to eliminate the oxygen and moisture in the pipeline. Then it enters the "*Box body clean*" state, as shown in **Figure 8(b)**, which is the screen to perform the box body cleaning. Then it waits for the T8 time, enters the next state of "*Detect body vacuum pressure*", starts to measure the vacuum pressure PT-1 (representing the body pressure), and waits for the pressure to reach the set vacuum level of the box body. When the vacuum pressure of PT-1 is reached, it will enter the "*Reach body vacuum pressure*" state. At this time, the EV-3 shutdown will be started first, the time will be waited for T2, then the P-1 will be closed, then the "*Detect box positive pressure*" state is entered, the glove box air supply replenishment valve EV-2 will be started, and N_2 gas will be added. When the positive pressure of the box body detected is sufficient, the system will enter the state of "*Body positive pressure adequate*" and close the EV-2 to stop the replenishment of the N_2 gas source and completes the box body cleaning work. The system will check if the set number of body cleaning has been reached. If it has not, it will return to the "*Box body clean*" state and start the next round of box cleaning until the set number of box cleaning has been completed. Then it will close the EV-3, CV1A, CV2A, CV1B, and CV2B valves and end the entire box body cleaning work.

3.4 Vacuum preservation and end preservation operations

The vacuum preservation operation is used to start the procedure for storing sensitive materials in the glove box. To perform a vacuum preservation operation, the user can activate the vacuum preservation control process after the material to be stored is placed in the glove box. The system first opens the box body vacuum valve EV-3 and activates the vacuum pump P-1 and simultaneously turns on the inlet valves of the regeneration system, CV1A, CV2A, CV1B, and CV2B to remove the oxygen and water in the pipeline, and then enters the "*Vacuum preserve*" state. The execution screen for vacuum preservation is similar to the body cleaning screen, as shown in **Figure 8(a)**, which is the screen to perform vacuum pumping. Then it waits for the T8 time, then enters the "*Detect body vacuum pressure*" state, and starts to measure the vacuum pressure PT-1 to wait for this pressure to reach the set box body vacuum level. Then it enters the "*Reach body vacuum pressure*" state. At this time, the EV-3 is turned off, and the T2 timing is started, when the vacuum

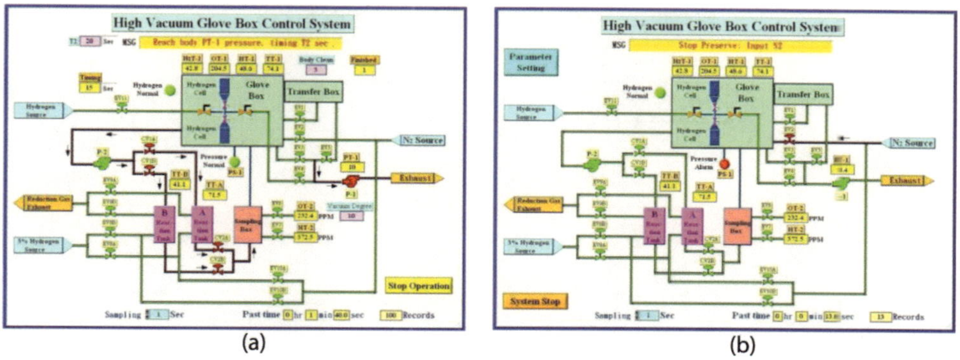

Figure 9.
Execution screen for vacuum preservation and end preservation. (a) Closing the vacuum valve of the box body at the vacuum preservation procedure and (b) supplementing the box body with positive pressure at the end preservation procedure.

pump P-1 continues to run, waiting for the EV-3 to complete the closing action, as shown in **Figure 9(a)**, which is the screen where the EV-3 is turned off first, and the P-1 continues to run. When the T2 time expires, the system will close the valves of CV1A, CV2A, CV1B, and CV2B. At this time, the box body can maintain a vacuum environment and the material storage operation can begin.

The end preservation operation is used to start the procedure for ending the material storage. When the user finishes the preservation operation, the system will first enter the "*End preserve*" state, open the glove box gas source replenishment valve EV-2, and start to input the N_2 gas to supplement the pressure of the box body and detect the box positive pressure PS-1, as shown in **Figure 9(b)**, which is the screen to supplement the box body positive pressure. When the detected positive pressure in the box body is sufficient, the system will first turn off the EV-2, stop the replenishment of the N2 gas source, then enter the "*Transfer box balance*" state, and the transfer box balancing valve EV-1 will be turned on. The gas in the body is introduced into the transfer box to bring it into pressure balance, then the time is counted to perform T1 time, and then the system enters the "*Exit preserve*" state. At this point, the user can open the inside and outside doors of the transfer box to take out the stored material.

3.5 Hydrogen cell cleaning and hydrogen storage operations

The hydrogen cell cleaning operation is used to remove the gas in the hydrogen storage cell to reach the set vacuum level, so as to facilitate the subsequent operation of storing and releasing hydrogen. When this operation is performed, the system will first turn on the hydrogen cell vacuum valve EV-4 and start the vacuum pump P-1 to enter the "*Hydrogen cell clean*" state, and start the T8 timing, as shown in **Figure 10(a)**, which is the hydrogen cell cleaning screen. After waiting for the T8 timer to complete, the system will enter the "*Detect cell vacuum pressure*" state. At this time, the pressure PT-1 (representing the vacuum pressure of the hydrogen cell) will be measured. It is necessary to wait for the pressure to reach the set vacuum level. When the pressure of PT-1 reaches the set vacuum level, the system enters the "*Reach cell vacuum pressure*" state. At this time, the closing action of EV-4 is started first, and the time is waited for T2 time. When the time of T2 is reached, the hydrogen cell cleaning operation can be ended.

Hydrogen storage operation can store and release hydrogen in the hydrogen cell in the glove box. When this operation is performed, the system first enters the "*Store*

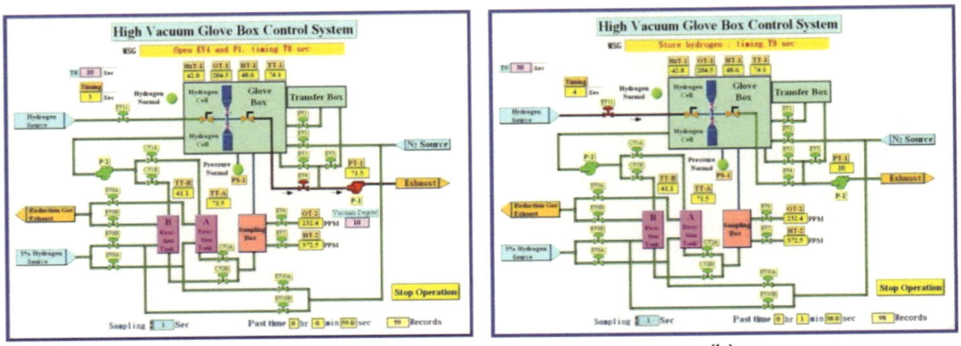

(a) (b)

Figure 10.
Execution screen for hydrogen storage tank cleaning and hydrogen storage. (a) Hydrogen cell cleaning procedure and (b) hydrogen storage procedure.

hydrogen" state, opens the hydrogen storage shut-off valve EV-11, and inputs hydrogen into the hydrogen cell of the glove box, as shown in **Figure 10(b)**, which is the initial screen for hydrogen storage. Then waits T9 time to finish the hydrogen storage operation and enters the "*Store complete*" state. In this process of hydrogen storage, if the hydrogen concentration is detected too high, the green "Hydrogen normal" light (as shown in **Figure 10(b)**) will be changed to flashing "Hydrogen alarm" red light, and the hydrogen storage operation will be immediately stopped.

4. Cycling cleaning regeneration control process

The glove box system contains two reaction tanks A and B. The cycling cleaning regeneration function can cycle between the two reaction tanks. When one reaction tank loses the cleaning capacity due to the run out of reagent, it can switch automatically to another reaction tank to work, and the reaction tank that losing the cleaning capacity performs the regeneration operation at the same time before joining the cycling work after they resume the cleaning capacity. In other words, this system can perform the cycle cleaning and cycle regeneration between reaction tanks A and B synchronously. The synchronization process is shown in **Figure 11**. In the figure, T3–T7 are the time parameters set by the user in **Figure 6(a)**, where T3 is the cumulative maximum working time of the reaction tanks (and the tanks can no longer work beyond this time), T4 is the time to regenerate the reagent by introducing 3% hydrogen, T5 is the time for the introduction of nitrogen to remove oxygen and water, T6 and T7 are the working time and the rest time, respectively, of the reaction tanks for each operation.

Figure 12 shows the state transition diagram of the glove box cycling cleaning regeneration control process. The dashed circles (states) and the arrow lines represent the parts that can be executed synchronously. Reaction tank A is preset to be the tank for the first time execution by the system, and the reaction tank B is in standby state.

The control system first enters the "*A works*" state, executes the cycling cleaning procedure of reaction tank A, it opens the cycling inlet valve CV1A of tank A, the cycling outlet valve CV2A, the oxygen sampling valve EV-6, the humidity sampling valve EV-7, etc., opens the clean pump P-2, and starts the timing of T3 and T6 at the same time, as shown in **Figure 13(a)**, which is the screen for tank A to perform the cleaning. At this point, the clean pump P-2 starts to send the oxygen and moisture in the glove box continuously to the reaction tank A for adsorption, so as to achieve

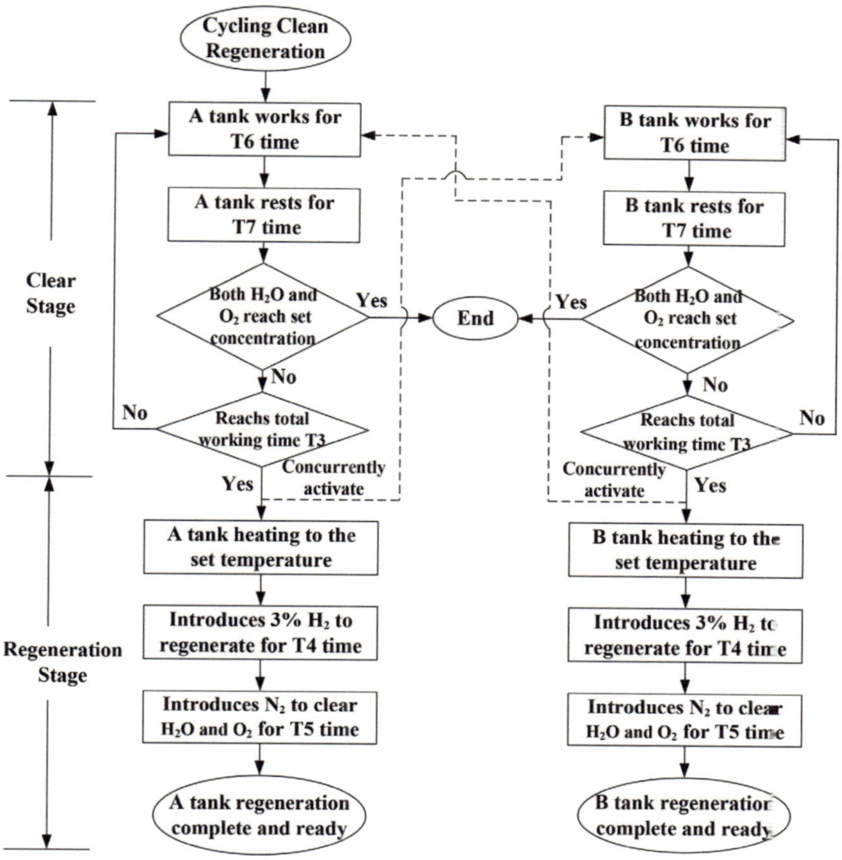

Figure 11.
Synchronization process flow of cycling cleaning regeneration.

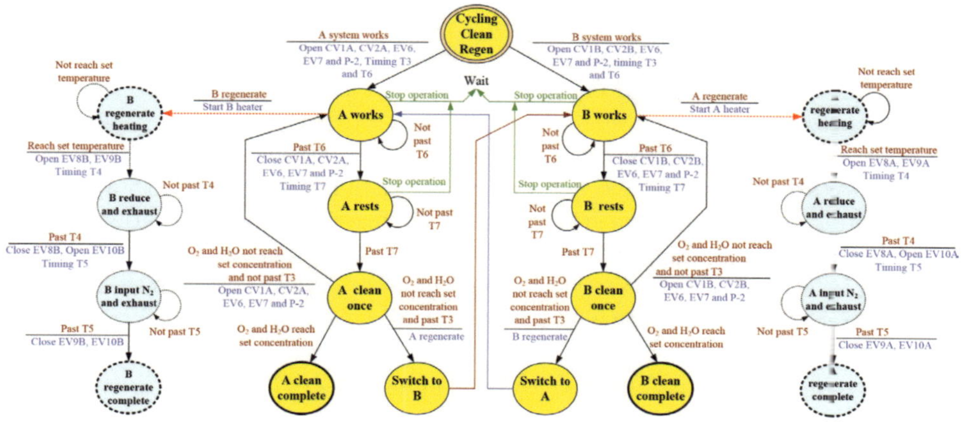

Figure 12.
State transition diagram of cycling cleaning regeneration control process.

the purpose of removing oxygen and moisture. When the T6 time is completed, the system enters the "*A rests*" state. At this time, the valves CV1A, CV2A, EV-6, EV-7, and the clean pump P-2 are all closed, and the T7 timing for rest starts, as shown in **Figure 13(b)**, which is the screen of the reaction tank A taking rest for the time of T7.

When the T7 timing is completed, the system will enter the status of "*A clean once*". At this time, the system will check whether both the detected oxygen

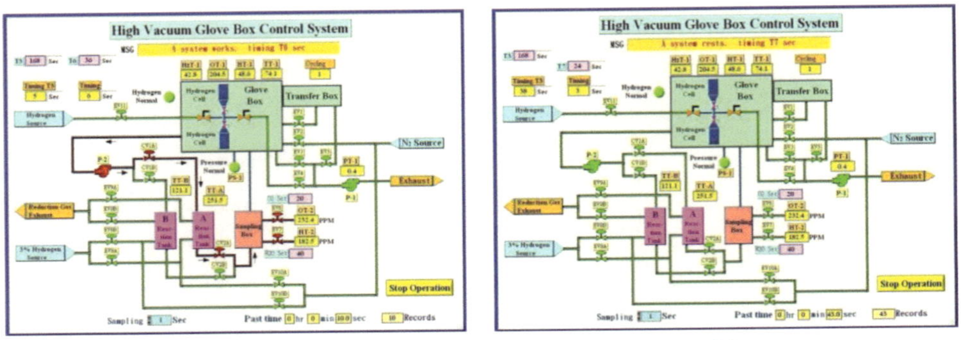

(a) (b)

Figure 13.
Screen for reaction tank A to perform cleaning operation. (a) Cleaning operation of reaction tank A and (b) reaction tank A taking rest for T7 time.

concentration OT-2 and the moisture concentration HT-2 have reached the set concentrations. If yes, the system will enter "*A clean complete*" state and automatically return to the "*Wait*" state of **Figure 6**. If OT-2 and HT-2 cannot reach the set concentrations at the same time, the system will check whether the cumulative working time T3 has been reached. If the T3 time has not yet been reached, it will return to the "*A works*" state and continue to perform the cycling procedure of tank A working first for T6 time and then resting for T7 time.

If T3 time has been reached and OT-2 and HT-2 have not yet reached the set concentrations, it means that the reaction tank A has lost the cleaning ability due to the run out of the reagent, and the system will enter the state of "*Switch to B*". This state will start the reaction tank A regeneration function first, and then move to the "*B works*" state for reaction tank B to perform the cleaning work, while tank A performs the regeneration process at the same time, as shown in **Figure 13(a)**, which is the screen for tank B to work and tank A in regeneration. When entering the "*B works*" state, the system will first open the CV1B, CV2B, EV-6, and EV-7 valves and the clean pump P-2 and start the timing of T6 and T3. Then it performs the cycling cleaning procedure of working for T6 time and then resting for T7. This control process is similar to the monitoring process when tank A is working.

When tank B performs the cleaning operation, tank A executes the regeneration process synchronously, and the state transition diagram thereof is shown in the rightmost blue dotted state area of **Figure 12**. Reaction tank A first enters the "*A regenerate heating*" state. The system controls the heaters to heat the reaction tank A [25–27], and detects whether the temperature TT-A has reached the set regeneration temperature. When the reaction tank A's temperature TT-A reaches the set temperature, the system will enter the "*A reduce and exhaust*" state. At this time, the 3% hydrogen inlet valve EV-8A and exhaust valve EV-9A will be turned on for 3% hydrogen to enter the reaction tank A to regenerate its cleaning agent, for the reduced waste gas to be discharged via the EV-9A valve, and the timing of T4 starts at the same time, as shown in **Figure 14(a)**, which is the screen for tank B to perform the cleaning work (which is now in the T7 rest phase) and for tank A to enter 3% hydrogen to regenerate its cleaning reagent.

When the T4 time expires, the system will enter the "*A inputs N_2 and exhaust*" state. At this time, the 3% hydrogen gas inlet valve EV-8A will be closed first, but the cleaning valve EV-10A will be opened to introduce dry N_2 gas in the gas source, so that the oxygen and moisture in the reaction tank A are taken away and discharged through the outlet valve EV-9A. And the timing of T5 starts, as shown in

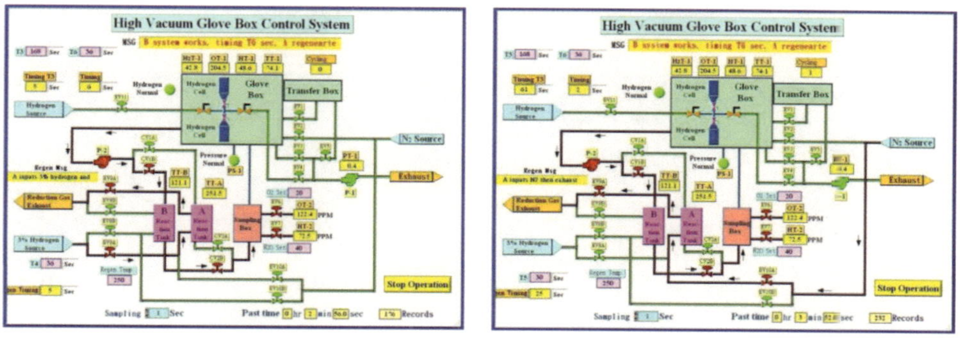

(a) (b)

Figure 14.
Screen of tank B performing cleaning operation and tank A in regeneration. (a) Tank B resting for T7 and tank A entering 3% hydrogen to regenerate the reagent and (b) tank B operating and tank A introducing dry gas to take away oxygen and moisture.

Figure 14(b), which is the screen where tank B operates and tank A introduces dry gas to take away the oxygen and moisture.

When the T5 time expires, the system enters the "*A regenerate complete*" state. At this time, EV-9A and EV-10A are turned off to end tank A regeneration operation. When tank B has several cycling of operation by working for T6 time and resting for T7 time several times and the detected oxygen and water concentration still fail to reach the set values, this means the cleaning agent has also been used up. At this time, as long as the T3 time has been reached, it will also start tank B regeneration and then enter the state of "*Switch to A*", to give the cleaning work back to tank A, and tank B will perform the regeneration work.

5. Curve rendering analysis

5.1 Vacuum test curve rendering analysis

The system automatically records the detected input signals for archiving during the execution of various glove box operation functions. The detected input signals are shown in the screen of **Figure** 5. By using the internal temperature TT-1, the primary humidity HT-1, the primary oxygen concentration OT-1, the vacuum pressure PT-1, and the PT-1 set value, the system can draw the vacuum test curves of the operation functions such as the transfer box cleaning, box body cleaning, and vacuum preservation, as shown in **Figure 15**, which is the elastic display screen of the vacuum test curve corresponding to the box body cleaning. **Figure 15(a)** shows the complete vacuum test curves. There are five switch buttons and five curve color setting boxes of "TT-1," "HT-1," "OT-1," "PT-1," and "PT-1 setup" on the left of the screen. Each switch button can be used to switch between "show/hide." Each color setting box enables the user to set the display color of the corresponding curve. At present, all five buttons are in the "show" state. Therefore, the above five test curves are displayed in the vacuum test graph in the middle of the screen. All the curves are normalized so that the display range is from 0 to 100%. From **Figure 15(a)**, we can see that the rate of decrease of water vapor concentration (72 → 2%) is higher than the rate of decline of oxygen concentration (26 → 2%).

The "show/hide" status of each switch button is set appropriately, allowing the user to analyze the relationship between various curves. If you want to analyze the vacuum pressure and the change of the primary water vapor concentration in

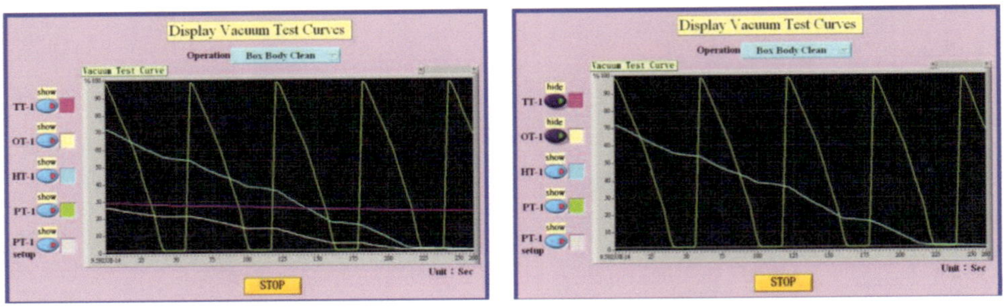

(a) (b)

Figure 15.
Display screen of vacuum test curves of box body cleaning. (a) Complete vacuum test curve and (b) change relationship between PT-1 and HT-1.

the box, you may click "TT-1" and "OT-1" switch buttons to set them to the "hide" status, and make the other three buttons "HT-1," "PT-1," and "PT-1 setup" maintain the "show" state, as shown in **Figure 15(b)**, which is the display of the relationship between the PT-1 and HT-1 curves. From this screen, it is observed that the green PT-1 pressure curve causes the significant drop of the blue HT-1 water vapor concentration during each descent (which represents the vacuumizing stage), and whether the PT-1 pressure curve can reach the gray PT-1 setup curve (i.e., reaching the setting value of the vacuum pressure) at each descent; and the pressure curve of PT-1 during the flat bottom is the nitrogen filling phase of the box body, when the change in the HT-1 water vapor concentration curve is less obvious. From the above analysis, we can see that the effect of the vacuumizing stage on the decline of water vapor concentration is obviously higher than that of the nitrogen filling stage.

5.2 Cleaning test curve rendering analysis

By using automatically recorded box temperature TT-1, regeneration temperature TT-A of the reaction tank A, regeneration temperature TT-B of the reaction tank B, the micro-water concentration HT-2 and micro-oxygen concentration OT-2 detected by the sampling tank, and the settings of HT-2 and OT-2, the system can draw the clean test curves of the cycling cleaning regeneration operation, as shown in **Figure 16**, which is the elastic display screen of the test curve of such cleaning. **Figure 16(a)** shows the complete cleaning test curves. There are eight switching buttons left side of the screen, such as "TT-1," "TT-A," "TT-B," "HT-2," "OT-2," "Regen Temp setup," "OT-2 setup," and "HT-2 setup" and eight corresponding curve color setting blocks. Each switching button can be used by the user to switch between show/hide, and each color setting box enables the user to set the display color of the corresponding curve. Currently, all eight buttons are in the "show" state, so the above eight test curves will be displayed in the vacuum test graph in the middle of the screen. The "show/hide" status of each switch button is set appropriately, allowing the user to analyze the relationship between various curve changes.

To analyze the relationship between the change of micro-oxygen concentration and the cycling cleaning regeneration phase, users can display five test curves such as TT-A, TT-B, OT-2, regeneration temperature setting, and OT-2 setting in this screen, as shown in **Figure 16(b)**. It can be seen from the figure that the reaction tank A performs cleaning for the T6 time (OT-2 curve drop) and rests for T7 time (OT-2 curve is flat). After two such cycles, the reaction tank B is immediately switched to perform such a cleaning rest cycle (the change in the OT-2 curve from continuing to fall till becoming flat), while the reaction tank A simultaneously

DOI: http://dx.doi.org/10.5772/intechopen.80423

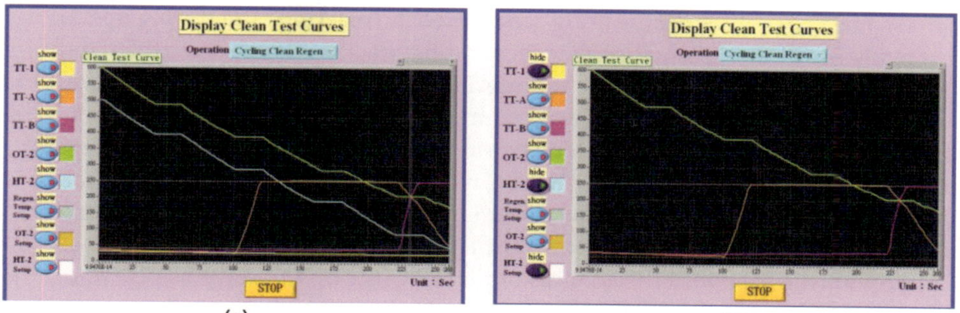

(a) (b)

Figure 16.
Display screen of cycling cleaning test curves. (a) Complete test curve of the cycling cleaning regeneration operation and (b) relationship among TT-A, TT-B, and OT-2.

starts the regeneration operation, that is, it is heated to the regeneration temperature (the orange TT-A curve rises to the regeneration temperature setting), and then 3% hydrogen is entered to regenerate the cleaning reagent. After the reaction tank B performs two cycles of cleaning rest, it will switch back to the reaction tank A to continue the cycling cleaning operation, when the reaction tank B will start the regenerating operation synchronously, because the red TT-B curve will rise due to heating the regeneration temperature setting. From the above changes in the curves, it can be seen that the operation of removing the micro-oxygen from the glove box can be performed in turn by two reaction tanks, and no interruption occurs due to the run out of the reagent of a reaction tank. In addition, for the change of the micro-oxygen concentration, performing two cleaning cycles in the tank A can reduce the micro-oxygen concentration from 1000 ppm to about 390 ppm. And two cycling cleaning operations by the reaction tank B is followed, and the micro-oxygen concentration can be further reduced to about 210 ppm.

6. Conclusions

This article develops an advanced control system for high vacuum deoxygenation and water removal for the glove box. This system can use the switching of the hardware pipeline and the control of monitoring software to provide users with the transfer box cleaning, box body cleaning, vacuum preservation, end preservation, hydrogen storage, hydrogen cell cleaning and cycling cleaning regeneration operating functions, and each operating function can be automatically executed through the settings of control parameters and instrument parameters. This system combined sequential control with multi-conditional logic control. In the state transition diagram of control process, there are mainly the timing control (according to the timing parameters of T1, T2, ... or T9) and multiple characteristic monitoring controls (such as temperature, vacuum pressure, oxygen concentration, water concentration, ... and other control parameters). The system provides users to set the target values of parameters for the sake of multi-conditional monitoring and logic control. Hence, the system may decide the direction of execution when facing with various events. As a result, the control output of multi-conditional judgment may be highly tolerant and stable toward signal offset and change in other external conditions. Therefore, this control system may be high robustness.

From the control point of view, the cycling cleaning regeneration function provided by the system must be considered in terms of its control, including the comparison between the detection values of water and oxygen concentration and the set values, whether the cleaning capacity of the reaction tank during operation is already insufficient, whether the regeneration reaction tank has completed the regeneration time, and so on. The monitoring host can use these judgment results to simultaneously perform the cycling cleaning and cycling regeneration control. This kind of control mode is a double loop control mode, in which cleaning and regeneration are performed at the same time. On the other hand, during the execution of various operating functions, the system can automatically record all measured data and display the vacuum test curves and cleaning test curves flexibly. This can provide the user with the analysis of performance on the glove box cleaning operation and vacuum operation, and the analysis results can be used as a basis for determining the quality parameters of the glove box and can also be used as a basis for the glove box user to plan experiments or operating procedures.

Acknowledgements

This study was sponsored by the project of Ministry of Science and Technology, R.O.C., project number: MOST 106-2221-E-344-004.

Author details

Ming-Sen Hu
Department of Aviation and Communication Electronics, Air Force Institute of Technology, Kaohsiung City, Taiwan, R.O.C.

*Address all correspondence to: mshu1227@gmail.com

References

[1] Ashby EC, Schwartz RD. A glove BOX system for the manipulation of air sensitive compounds. Journal of Chemical Education. 1974;**51**(1):65-68

[2] Boylen CW. Installation and use of the microscope within a gastight glove Box. Journal of Applied Microbiology. 1971;**21**(6):1089-1090

[3] Cournoyer ME, Castro JM, Lee MB, Lawton CM, Ho Park Y, Lee R, Schreiber S. Elements of a glovebox glove integrity program. Journal of Chemical Health and Safety. 2009; **16**(1):4-10

[4] Dryburgh PM. A glove-box and drying system for the manipulation of moisture sensitive materials. Journal of Scientific Instruments. 1967;**44**(8):658-660

[5] Smith C, Mullins D, Nakaoka R. A novel process control system for a glove box vitrification system. In: WM'04 Conference. 2004

[6] Su CS. Vacuum Technology Essence. Taipei, Taiwan, R.O.C: Wu Nan Book Corp.; 2013. ISBN13: 9789571134789

[7] Wu SC, Wang RW, Lin MY. Vacuum Technology and Applications. Taipei, Taiwan, R.O.C: San Ming Book Corp.; 2001. ISBN13: 9789570286762

[8] Cheng HB, Chu CJ. Painting Vacuum Technology. Taipei, Taiwan, R.O.C: Shu Chuan Publisher; 2012

[9] Campbell J. Speed Cleaning. New York: Rodale; 2005. p. 97. ISBN 1-59486-274-5

[10] National Laboratory Research Institute of Instrument Science and Technology. Vacuum Technology and Applications. New Taipei City, Taiwan, R.O.C: Chuan Yua Book Corp.; 2004. ISBN: 9570286768

[11] Cournoyer ME, Lee MB, Schreiber S. Minimizing glovebox glove failures, Part IV: Control charts, LA-UR 07-0783. Journal of the American Society of Mechanical Engineers. In: Proceeding from WM'07; 25 February-1 Mar; Tucson, Arizona. 2007

[12] Smith JD. The Virtual Glovebox (VGX): An immersive simulation system for training astronauts to perform glovebox experiments in space. Enclosure. 2002;**15**:8-13

[13] Hu MS. A high vacuum deoxygenation and water removal glove box control system. In: Aviation Technology and Fight Safety Conference; June 9; Haoshuing. 2017

[14] Glovebox Glove Improvement Workshop, LA-UR 06-5176. In: American Glovebox Society Annual Conference 2005; July 16-20; San Antonio, Texas. 2006

[15] Uematsu S, Kashiro K, Tobita N. Dismantling of gloveboxes for MOX fuel fabrication by a glovebox dismantling facility. In: WM'02 Conference; 24-28 February; Tucson, AZ. 2002

[16] Glove Box Workstations. MBRAUN. Available from: http://www.mbraun.com/products/glovebox-workstations

[17] Michael EC, Stephen S. Glove Testing & Evaluation, LAUR 07-3150, JOWOG 30 Exchange, Atomic Weapons Establishment; 21-25 May; Aldermaston, England. 2007

[18] Hu MS, Su RH. A glove box measuring control system and its measuring control method. Patent of the Republic of China, patent number: I594854, duration: 2/2017 to 1/2037

[19] Hu MS, Chen LH. The development of a high vacuum glove box control

system with cycling cleaning and regeneration. In: IEEE International Conference on Systems Science and Engineering; 21-23 July; Vietnam. 2017

[20] Advantech USB-4711A User Manual, 150 KS/s, 12-bit Multifunction USB Data Acquisition Module, Advantech

[21] USB-4750 User Manual, 32-Channel Isolated Digtial I/O USB Module, Advantech

[22] Hsiao TC, Wang CY, Chu JW. Virtual Instrument Control Programming—LabVIEW 8X. New Taipei City, Taiwan, R.O.C: Gao Li Book Corp.; 2014

[23] Chen CC. LabVIEW Applications (Include Automatic Measurement and Remote Control). New Taipei City, Taiwan, R.O.C: Chuan Yua Book Corp.; 2014

[24] Hsu YH. Interface Design and Practice-Using LabVIEW. New Taipei City, Taiwan, R.O.C: Chuan Yua Book Corp; 2014

[25] Goodwin GC, Graebe SF, Salgado ME. Control System Design. New Jersey: Prentice Hall; 2001

[26] Kuo BC. Automatic Control Systems. New York: Holt, Rinehart and Winston, Inc.; 2003

[27] Nise NS. Control System Engineering. 4th ed. New York: John Wiley & Sons, Inc.; 2004

Chapter 4

Artificial Intelligent-Based Predictive Control of Divided Wall Column

Rajeev Kumar Dohare

Abstract

Distillation is the most popular thermal separation technique used in the chemical and petrochemical process industry for the liquid mixture separation. Certainly, the distillation has a plenty of advantages, yet it has a drawback such as more energy requirement. In order to reduce the energy consumption of the conventional distillation column, energy integration is applied within the distillation column. The most important thermally coupled distillation column sequence is the Petlyuk column, which uses two recycle streams. Petlyuk column is a novel design that integrates two distillation columns into one shell, which is known as a dividing wall column. A dividing wall column (DWC) offers the possibility to separate a multicomponent mixture into high-purity products or sharp splits. Artificial neural network predictive controller (ANNPC) has been implemented to control the DWC. The performance of the ANNPC control strategies and the dynamic response of the DWC are investigated in terms of the product composition in the different sections of the dividing wall column for the different persistent disturbances such as feed flow rate, feed compositions, and liquid split factor.

Keywords: artificial intelligent, DWC, BTX, distillation, simulation, MATLAB

1. Introduction

Divided wall column is the combination of four thermally coupled distillation columns. The schematic diagram of the DWC is shown in **Figure 1**.

Benzene-toluene-o-xylene (BTX) system as feed is the dividing wall column for separation purpose. Feed introduced into the prefractionator side of the wall. A side stream is removed from the other side. The side stream is mostly the intermediate boiling component of the ternary mixture. The lightest component (benzene) goes overhead in the distillate product, and the heaviest component (o-xylene) goes out in the bottom product. Therefore, a single dividing wall column can separate a ternary mixture into three pure product streams. Due to high interactions among the process variables and due to restricted experiences, it is very difficult to control by the conventional controller.

In many chemical processes, artificial neural network has been implemented successfully in the process identification and control of the nonlinear dynamic systems. This is the main reason to switch over to the recent control strategies such as artificial neural network, fuzzy logic, etc. All such control technique-based

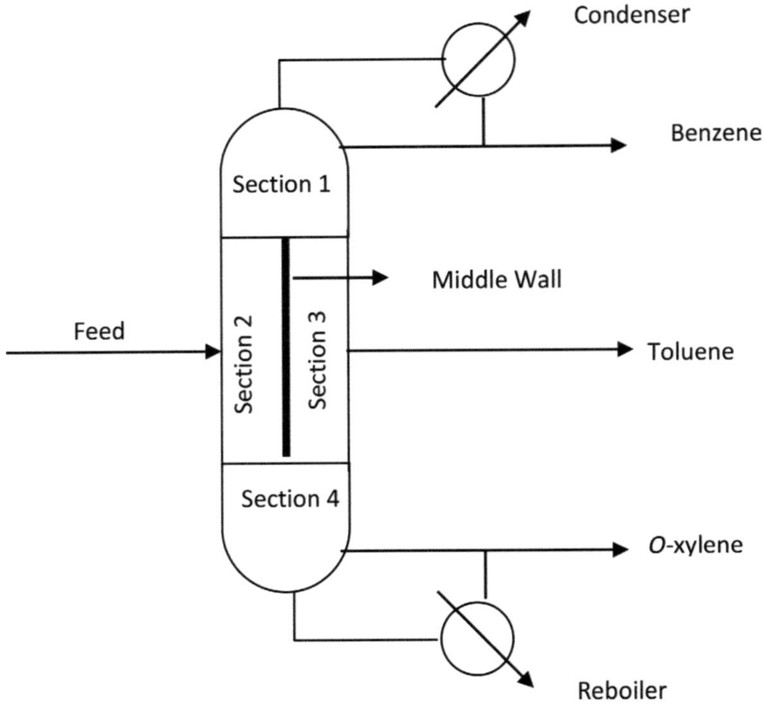

Figure 1.
Schematic diagram of divided wall distillation column.

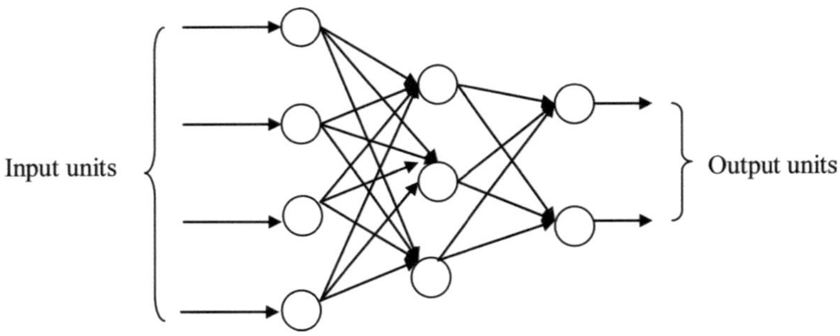

Figure 2.
ANN structure.

controllers are known as intelligent controllers. Settling time in the model-based controller is less in comparison to the conventional controller without compromising the product purities. To further improve the control performance, the artificial intelligence techniques were also attempted.

Artificial neural networks (ANN) have been designed on the complexities of the brain functions in an effort to capture the amazing learning capabilities of the brain as shown in the **Figure 2**. ANN is a parallel computer or processor designed to imitate the way the brain accomplishes a certain task [1]. The smallest processing element of ANN is a neuron or node, which helps to do sample calculations. Using the neurons collectively with massive connections among them results in a network that is able to process and store relative information for mapping the network inputs to its outputs. With this feasibility and its capability, there are most widespread interests in solving complicate problems particularly in the fields of pattern recognition, control, forecasting, classification, system identification, and optimization.

Authors already discussed, about the desired product purity at the optimized operating parameters viz. feed tray location, reflux ratio, liquid reflux rate, vapor split ratio, etc. [2]. To control the dividing wall column is the big challenge for the control engineers due to its high dynamics. Conventional controller like PID is not performed well in comparison to the advanced controller which was shown by the authors in their research [3]. Artificial neural network has been applied on dividing wall column for controlling the tray temperatures that is based on predictive technique. Composition of the product in the dividing wall column is the main controlled variable but due to the measurement delay in the online analyzer, it is rarely used. For successful operation, monitoring, and controlling chemical process, an accurate online analyzer of important quality variables is essential. However, the measurement of all these variables online is a big task due to the limitations such as the high cost, time delay, and reliability, therefore they cannot be directly close-loop controlled. Composition is controlled indirectly by controlling the temperature of the different sections in DWC. To enhance the controllability, artificial neural network-based predictive controller (ANNPC) can be used to control the temperature of the different sections in the dividing wall column.

In actual industrial circumstances, it would be ensured that the equipment should be run safely and efficiently with the important process variables that relate mostly with system stabilizations and product qualities have to be controlled in real time. However, it is very difficult to measure these variables by online physical sensors and its economic issue [4]. Due to the importance of the control problem, many methods have been adopted in the past: the first method is a quality open-loop control, in which to get the quality of the products, the undue purification is required; the second method is useful indirectly to control the quality of close-loop, such as controlling the temperature at each plate in the column at different sections. But sometimes, this method cannot control the quality of the product; third method is online process gas chromatography. Due to its limitations like cost, reliability, and the time delay, it cannot satisfy the online control requirement of quality.

2. MIMO neural network generation

The artificial neural network predictive control was trained using Levenberg-Marquardt optimization algorithm. Hidden neurons use the sigmoid activation function, whereas output layer neurons use the linear activation function. The selection of the different layer neurons depends on the complexity of the problem. On a trial-and-error basis, a number of neurons were selected in this present study. The general structure of the neural network is shown in **Figure 1**. For the normalization, all the inputs and outputs should be dealt up with different magnitudes. The total number of data is divided into different parts for neural network model building: (1) training data, (2) validation data, and (3) testing data. First part of the data is used for training purpose, the second part for validation of the network structure, and third part of the data is useful to evaluate the selected model.

3. Artificial neural network predictive controller

Artificial neural network predictive control (ANNPC) is a combination of artificial neural network- and model-based controller. The multilayer

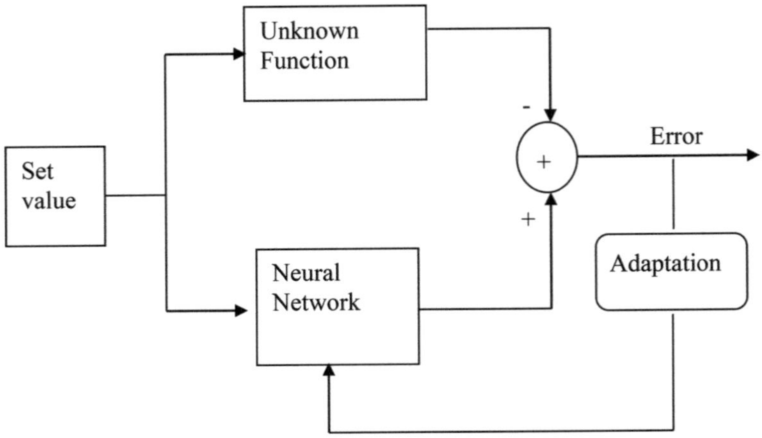

Figure 3.
Line diagram of neural network predictive control.

perceptron makes it a good choice for modeling nonlinear systems and for implementing nonlinear controller. The line diagram for the use of a neural network implemented on a process is shown in **Figure 3**. The unknown function may correspond to a controlled system, and the neural network is the identified plant model. Two-layer networks, with sigmoid transfer functions in the hidden layer and linear transfer functions in the output layer, are universal approximations. Artificial neural network is considered here as a predictive model for controlling the BTX system. The ultimate achievement of the model-based predictive controller is to generate a sequence of control signals minimizing a cost index that is the function of difference between the future process outputs from the desired set points and control moves. The ANN control structure was designed in "nntool" box, which was exported in Simulink environment to control DWC. An S-function was written in MATLAB for representing the DWC model. This DWC model was integrated with neural network control structure in MATLAB.

4. Data generation for training, testing, and validation of the network

To generate training data for the neural network training, a uniform random number generator for all the three input variables has been taken in the Simulink model as shown in **Figure 4**. The random number was kept constant at least for 1000 s. About 700 samples of each variable were generated with a sampling period of 100 s. These entire sample data were divided into three segments, viz.: training data, validation data, and testing data, in the ratio of 70:15:15, respectively.

The output performance of neural network depends on the number of neurons in input layer, number of hidden layer and optimizing algorithms. For the better output performance of the neural network, 90 neurons have been chosen in the input layer. The 90-input neurons correspond to 15 past values of each input and output variables. The numbers of neurons in hidden layer were found by the performance index such as R2. The performance analysis of the neural network is shown in **Table 1**.

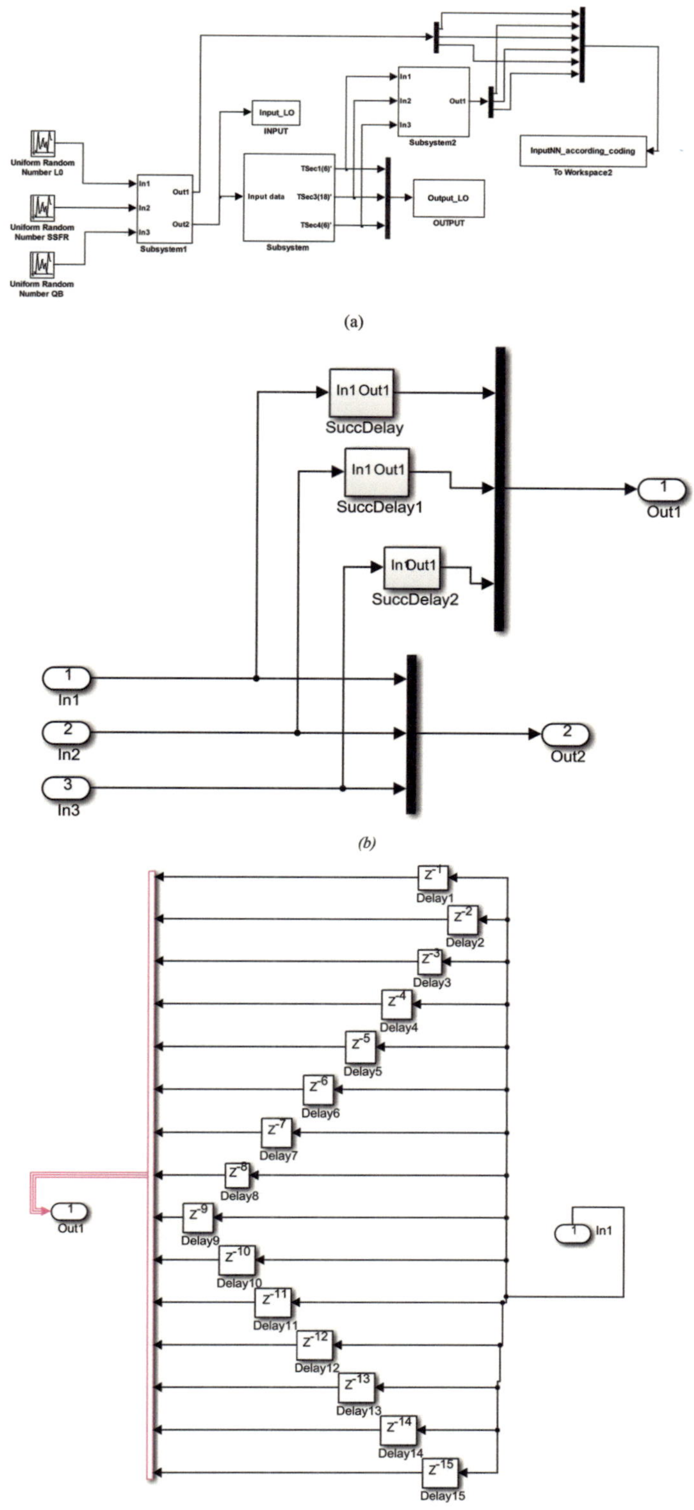

Figure 4.
Systematic Simulink model of the dynamic DWC for data generation. (a) Main figure, (b) subsystem1, and (c) subsystem of subsystem1 "SuccDelay."

Training algorithm	Performance function	Neurons in hidden layer	Epoch	Max validation check	Tolerance	Training	Validation	Testing
Levenberg-Marquardt backpropagation	Mean squared normalized error	10	200	100	1e-7	0.93	0.95	0.96
		10	150	100	1e-7	0.96	0.88	0.93
		10	100	100	1e-7	0.94	0.92	0.95
		10	100	100	1e-6	0.95	0.91	0.94
		10	50	25	1e-6	0.95	0.94	0.93
		10	50	25	1e-7	0.93	0.97	0.98
		10	20	10	1e-7	0.94	0.96	0.97

Table 1.
Performance parameters of neural network.

5. Neural network training and its algorithm

Delgado et al. [5] suggested controlling the process online by the neural network; it requires accurate network training data to design some network frames, so that the model has better extrapolation and suaveness ability [5].

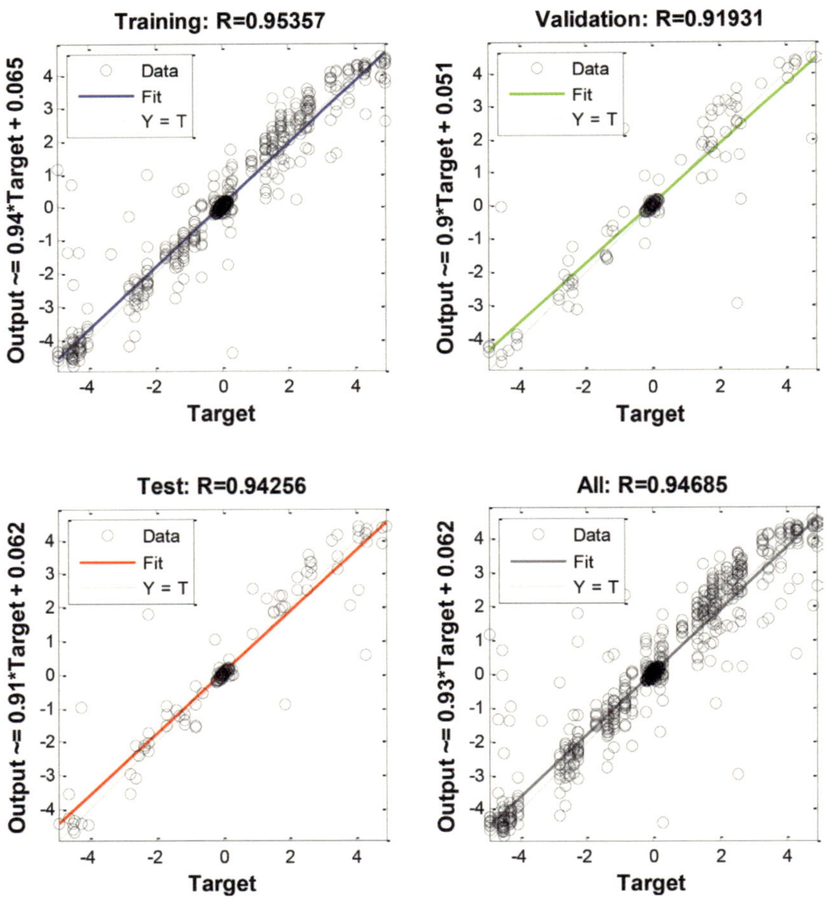

Figure 5.
Training, testing, and validation of the generated data of DWC.

When artificial neural network technique is applied on any process model, the common method is to collect the data for training either directly from the process plant or from the simulation. The training datasets for the network generation in the neural network have been obtained by the open-loop process under dynamic operating conditions because all the steady-state variables are in nonoscillatory motion. Due to nonavailability of the experimental BTX data of dividing wall distillation column, a mathematical model has been developed and then simulated in real time to find out the training data for network generation. Backpropagation algorithm has been used as a training algorithm to tune the connection weights for the function of each neuron.

For the offline training of the neural network, different types of training algorithms such as Levenberg-Marquardt, gradient-descent, and conjugate gradient can be used. But Levenberg-Marquardt algorithm is a fast-converging optimization technique among all. For DWC, the training, testing, and validation results are shown in **Figure 5**. The results show that there is a good fit between actual data and predicted data.

This algorithm works on the gradient and Hessian matrix of the objective function. In the Levenberg-Marquardt algorithm, Hessian matrix is approximated by Jacobian matrix and it is calculated by second-order derivative of objective function [6]. It is an easy way to improve the calculation up to the desired precision.

6. Simulink model of artificial neural network predictive controller

A Simulink model was designed to analyze the control performance of the artificial neural network predictive controller on the various load disturbance parameters. A systematic block diagram of the artificial neural network predictive controller is given in **Figure 6**.

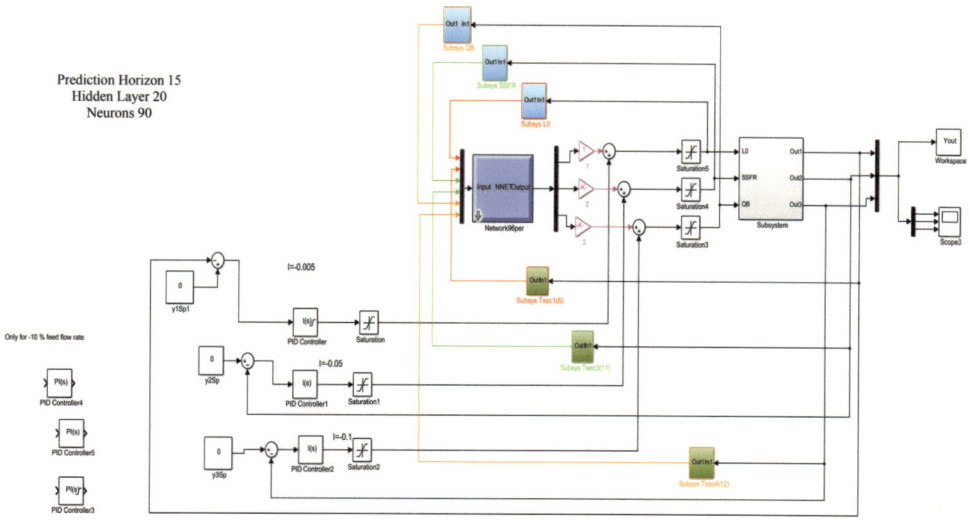

Figure 6.
Systematic block diagram of artificial neural network predictive control.

7. Results and discussions

7.1 Effect of benzene composition change in feed

Load change of ±10% in benzene composition of the feed was given to check the performance of ANNPC as shown in **Figure 7**. The controller was able to bring the temperatures in all the sections to the desired set points without any offset. The settling time in section 1 and section 4 for ±10% change in benzene composition is nearly 0.6 and 0.45 h, respectively. Although main column has 0.32 and 0.42 h settling time for −10% change and +10% change in benzene composition, respectively. The maximum peak of temperature in the section one (Tsec1) is 0.4 and 0.28°C at −10% change and +10% change, respectively. Section 3 temperature (Tsec3) and section 4 temperature (Tsec4) have very small peak value in the range of 10^{-3} and 10^{-1}°C for the ±10% change. With respect to the temperature change of the section 1, benzene composition reaches to the desired set point without any offset in 0.6 and 0.7 h for −10% and +10% change. Moreover, toluene and xylene composition acquires the desired value in 0.5 and 0.4 h, respectively without showing moderate offset and small peaks (i.e., in the range of 10^{-4} and 10^{-5}). Meanwhile, the entire manipulated variable varies independently to control the composition indirectly.

Load change of ±20% in benzene composition of the feed was also given to check the robustness of ANNPC as shown in **Figure 8**. In this case also, the controller was able to bring the temperatures in all the sections to the desired set points without any offset.

7.2 Effect of toluene composition change in feed

Figure 9 shows the effect of ±10% change in toluene composition of the feed as a load change in the system. Temperature overshoot in the section 1 at +10% change is 0.81°C and at −10% change is 0.41°C. The rest two sections have overshoot of 0.01 and 0.05°C, respectively. The temperatures acquire the desired set point without showing any offset in sections 1, 3, and 4 up to 0.8, 0.7 and 0.5 h, respectively. Corresponding to temperature variation in the individual section, composition also varies. At ±10% change in toluene composition, benzene composition shows the maximum peak in comparison to the toluene and xylene compositions. Due to high peak in the benzene composition, the settling time is also more (i.e., 0.66 h) with respect to the other two composition settling times. All the compositions in the different sections are also controlled with marginal offset. Load change of ±20% in

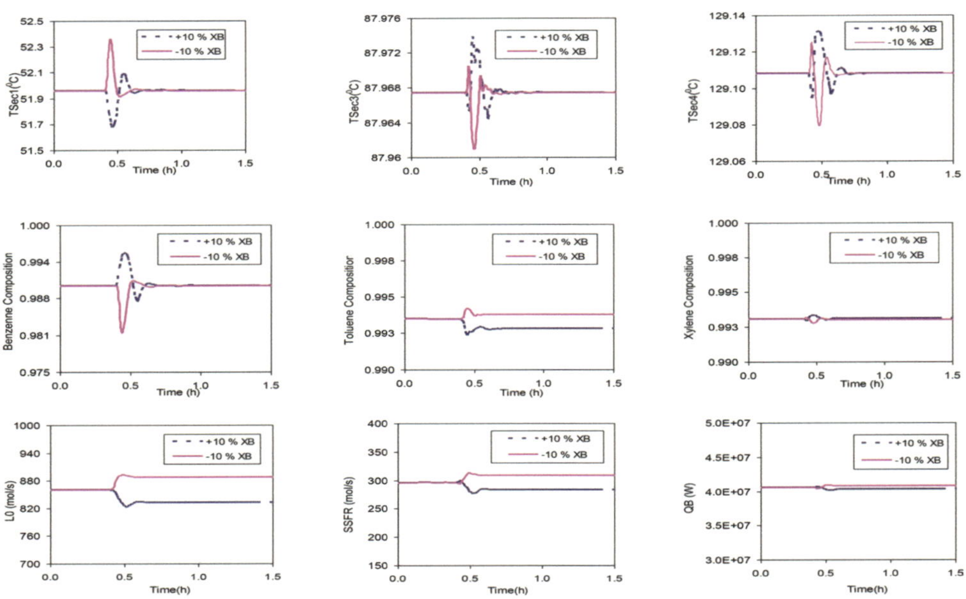

Figure 7.
±10% change in benzene composition in feed.

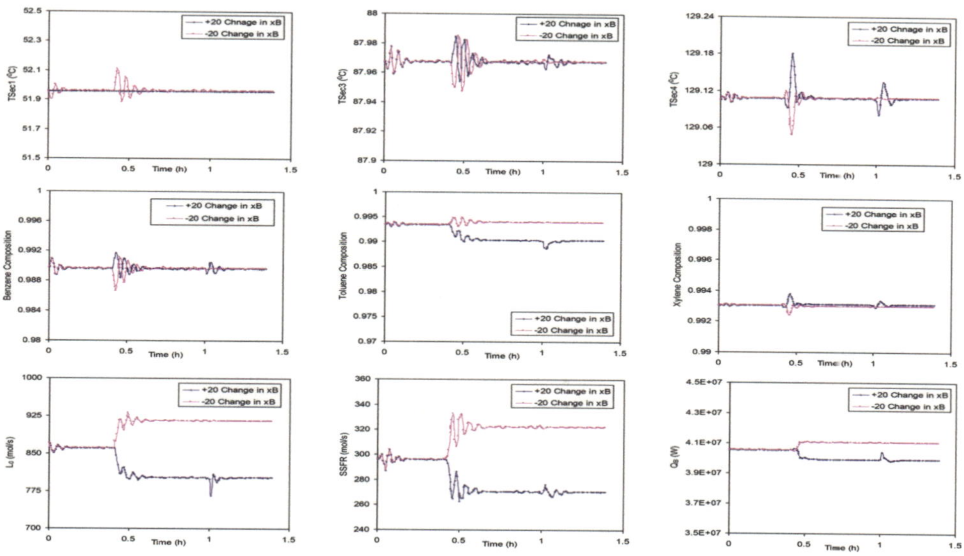

Figure 8.
±20% change in benzene composition in feed.

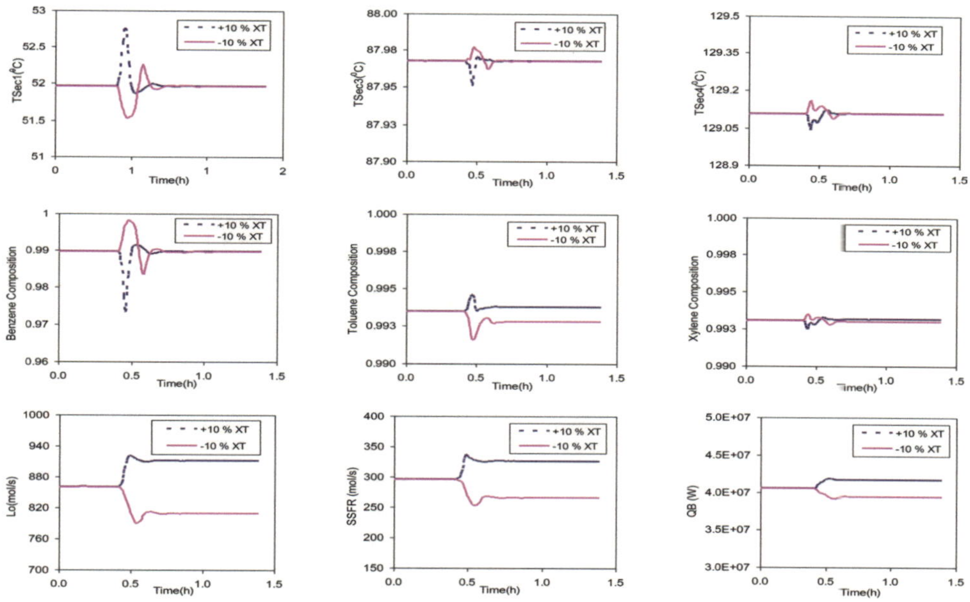

Figure 9.
±10% change in toluene composition in feed.

toluene composition of the feed was also given to check the robustness of ANNPC as shown in **Figure 10**. In this case also, the controller was able to bring the temperatures in all the sections to the desired set points without any offset.

7.3 Effect of o-xylene composition change in feed

A load change of ±10% in o-xylene composition was also imposed to check the performance of the controller as shown in **Figure 11**. As a ±10% change in o-xylene composition, temperature overshoot in the last section is near to 0.06°C. In comparison to the stripping section, rectifying section shows the maximum peak and

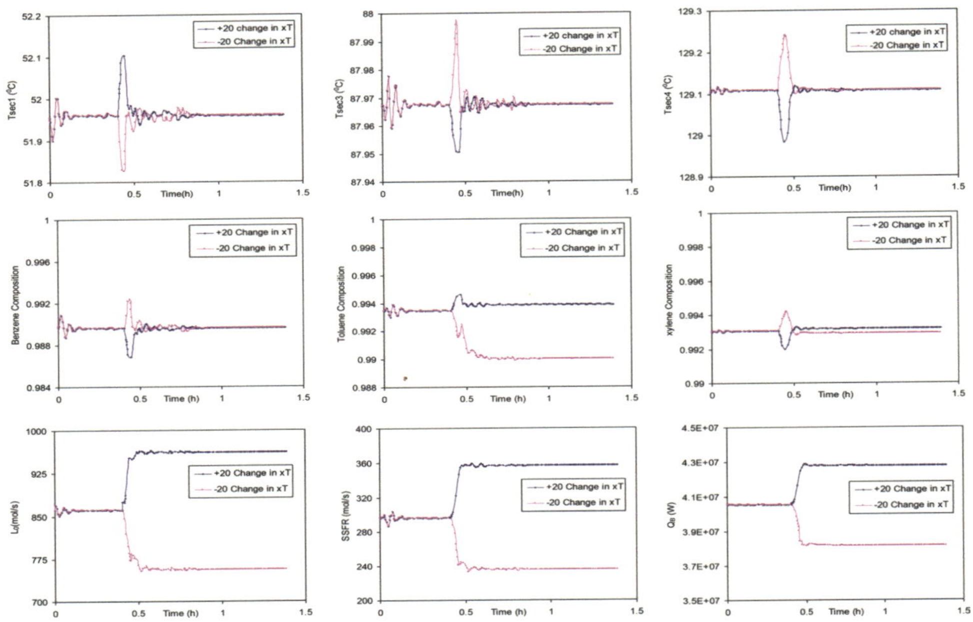

Figure 10.
±20% change in toluene composition in feed.

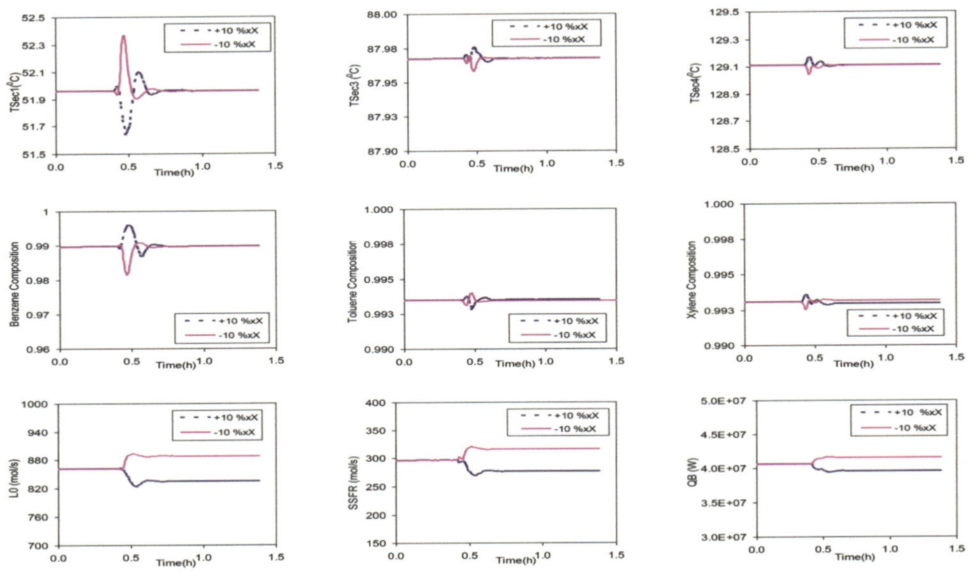

Figure 11.
±10% change in xylene composition in feed.

the main column shows minimum peak. After giving a load change of ±10% in o-xylene composition, benzene composition gets settled in 0.32 h, while toluene and o-xylene compositions take only 0.22 h to achieve the steady-state condition. Moreover, all the composition peaks are in the range of 1×10^{-3} (mole fraction), which is very small. Settling time of the temperature profile is 0.5 h in the rectifying section; 0.3 h in the main column and stripping section. Composition profiles of the sections 3 and 4 show very small offset and section 1 does not have any offset. The controller also showed the robustness for ±20% change in xylene composition in the feed as shown in **Figure 12**.

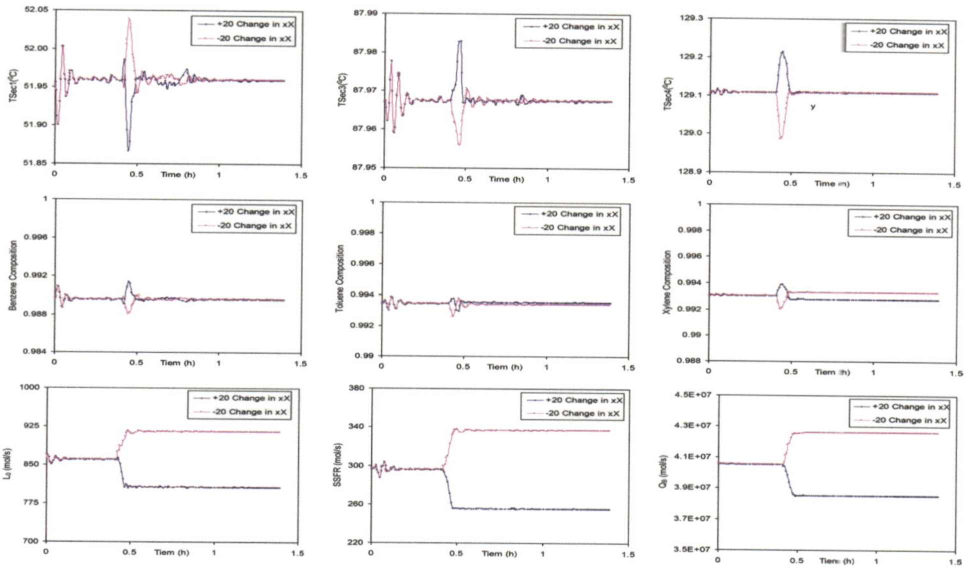

Figure 12.
±20% change in xylene composition in feed.

7.4 Load change in feed low rate

About ±10% load change was given in the feed flow rate to see the performance of ANNPC as given in **Figure 13**. The temperatures in all the three sections are brought back to the set point without significant offset. The temperature overshoot is very high in section 1 at +10% change in feed flow rate in comparison to −10% change. Moreover, sections 3 and four have similar peaks in both sides. Due to the high temperature peak in section 1 at +10% change, the settling time is twice of that in −10% change.

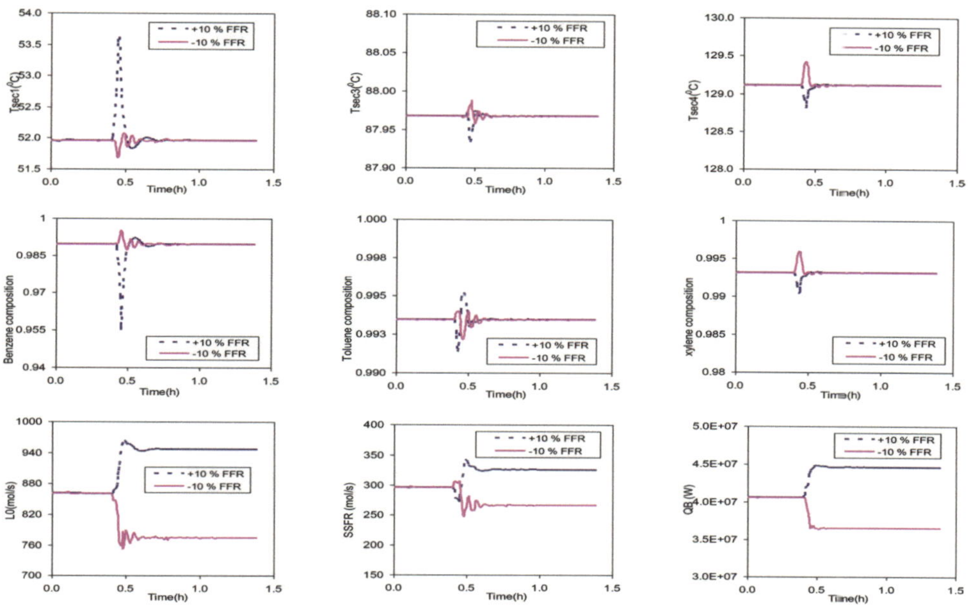

Figure 13.
±10% change in feed flow rate.

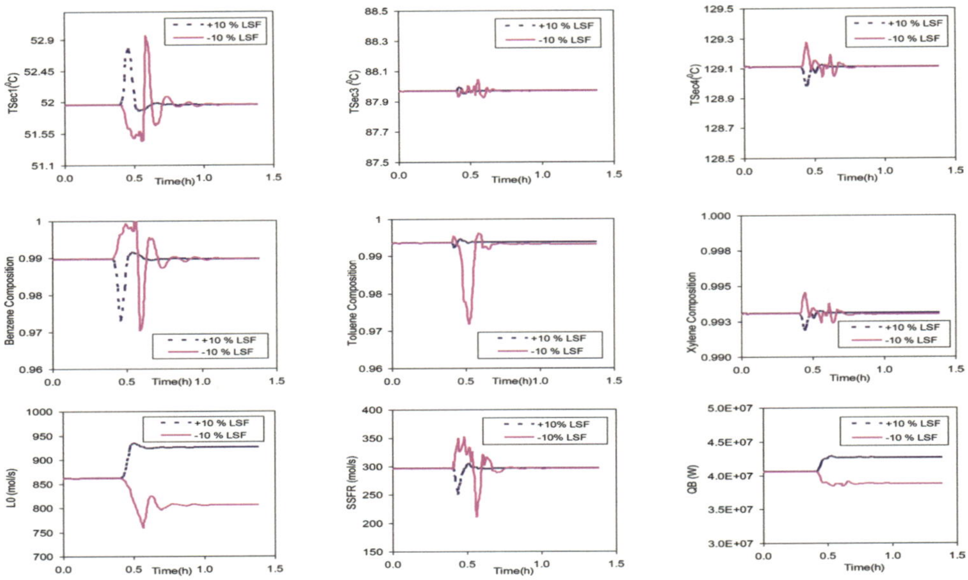

Figure 14.
±10% change in liquid split faction.

Change in feed flow rate creates more disturbances in the main column of the DWC, and therefore, the settling time in this section is more in comparison to the others. The peak of benzene composition at +10% change is 0.0346°C and −10% change is 5.4×10^{-3}°C, which is too low as compared to +10% change in feed flow rate.

7.5 Load change in liquid split factor

To analyze the performance of the ANNPC, liquid split factor was also considered as load disturbance as shown in **Figure 14**. Due to liquid distribution in the main column and the prefractionator column, variation in temperature profile is more. Settling time in section 1 is more in comparison to the other load changes. The overshoot in this section is 0.017 (mole fraction) at +10% change and 0.019 (mole fraction) at −10% change. Overshoot in the section 3 is 0.016 (mole fraction) only at −10% change. Moreover, o-xylene composition overshoot in section 4 is in the range of 1×10^{-3} (mole fraction).

8. Conclusion

To analyze the control performance of artificial neural network, random training data were generated. In total, 700 data were generated on the time interval of 100 s. Generated data were divided into the ratio of 70:15:15 for training, testing, and validation, respectively. Architecture of the neural network was assumed to consist of 90 input and 3 output neurons, which showed R2 of 96% for validation. The load changes in feed flow rate, feed composition, and liquid split factor were also induced to find the control performance in all the three sections by manipulating variables, viz. reflux rate, side stream flow rate, and reboiler heat duty. Settling time in the ANNPC controller was found to be very low in comparison to conventional controller. The ANNPC was also tested for ±20% change in feed composition of benzene, toluene, and xylene, which confirmed the robustness of this controller with respect to change in feed composition.

Author details

Rajeev Kumar Dohare
Malaviya National Institute of Technology, Jaipur, India

*Address all correspondence to: rajeevdohare@gmail.com

References

[1] Willis MJ, Di Massimo C, Montague GA, Tham MT, Morris AJ. Artificial neural networks in process engineering. IEE Proceedings D Control Theory and Applications. IET. 1991;**138**(3):256-266

[2] Dohare RK, Singh K, Kumar R, Upadhyaya SG, Gupta S. Dynamic model of dividing wall column for separation of ternary system. In: Chemca-2011; Australia. 2011

[3] Dohare R, Singh K, Kumar R. Modeling and model predictive control of dividing wall column for separation of BTX. Systems Science & Control Engineering. 2015;**3**(1):142-115

[4] Park S, Han C. A nonlinear soft sensor based on multivariate smoothing procedure for quality estimation in distillation columns. Computers & Chemical Engineering. 2000;**24**:871-877

[5] Delgado A, Kambhampati C, Warwick K. Dynamic recurrent neural network for system identification and control. IEE Proceedings-Control Theory and Applications. 1995;**142**:307-314

[6] Nogaard M, Ravn OE, Poulsen NK, Hansen LK. Neural Networks for Modelling and Control of Dynamic Systems: A Practitioner's Handbook. London: Springer-Verlag, Inc; 2000

Section 2

Control Techniques

Chapter 5

Modeling and Attitude Control of Satellites in Elliptical Orbits

Espen Oland

Abstract

The attitude determination and control system (ADCS) for spacecraft is responsible for determining its orientation using sensor measurements and then applying actuation forces to change the orientation. This chapter details the different components required for a complete attitude determination and control system for satellites moving in elliptical orbits. Specifically, the chapter details the orbital mechanics; perturbations; controller design; actuation methods such as thrusters, reaction wheels, and magnetic torquers; actuation modulation methods such as bang-bang, pulse-width modulation, and pulse-width pulse-frequency; as well as attitude determination using vector measurements combined with mathematical models. In sum, the work describes in a tutorial manner how to put everything together to enable the design of a complete satellite simulator.

Keywords: ADCS, attitude control, attitude estimation, thrusters, reaction wheels, magnetic torquers, elliptical orbits, PWM, PWPF, bang-bang, Madgwick filter, Sun vector model, magnetic field model, sliding surface control, quaternions, angular velocity

1. Introduction

The problem of developing attitude determination and control systems (ADCS) has received much attention in the last century with general books such as [1–4], as well as description of individual ADCS designs for different satellites in works such as [5–8]. While Refs. [1, 2] can be considered excellent foundation books within the topic of spacecraft modeling and control, there is a need for a more concise presentation of the attitude control problem and how this can be modeled in a simple manner, both not only as a tutorial for new researchers but also to give insight into the different components required for ADCS design drawing on ideas and results from previous works as mentioned above.

This chapter is an extension of [9] and builds on much of the previous work, as well as the research done through the HiNCube project as presented in [10–12]. This work considers the problem of designing a complete ADCS system comprising all the required components. **Figure 1** shows the control structure and the required signal paths, giving an overview of the contents in this chapter, as each block is described in detail to put the reader in position to design their own ADCS system. The required sensors for this system are a magnetometer to measure the Earth's magnetic field, a gyro to measure the angular velocity of the satellite, as well as Sun sensors for measuring the direction toward the Sun. Further, the mathematical

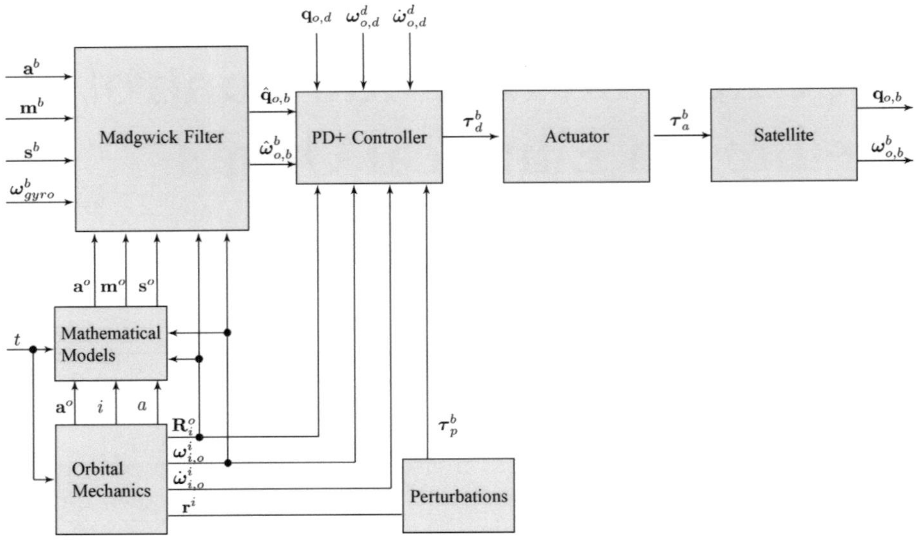

Figure 1.
This figure shows the different components required for modeling and controlling a satellite in an elliptical orbit and shows the main components required for creating a satellite simulator.

models used together with sensor measurements to determine the attitude of the satellite requires a real-time clock, as the time and date are required to know the direction toward the Sun as well as what the magnetic field looks like at a given day and time. Comparing the sensor measurements with the mathematical models allows for the determination of the attitude of the satellite, something that is done using the Madgwick filter as presented in [13]. With an estimated attitude obtained using the Madgwick filter, the attitude can be controlled to point a sensor onboard the satellite in a desired direction, something that is solved in this chapter using a PD+ attitude controller, calculating the desired torques required in order to make the attitude and angular velocity errors go to zero. In order to create the desired moments, this chapter presents how this can be achieved using a number of different actuators, namely, magnetic torquers, reaction wheels, and thrusters. The orbital mechanics block describes how the satellite moves in its elliptical orbit, while the perturbing forces and moments block describe how the different perturbations affect the satellite. Simulations show the performance of the different methods and should put the reader in a position to simulate and design new attitude determination and control systems for satellites in elliptical orbits.

2. Mathematical modeling

2.1 Notation

This subsection is similar to the author's previous works, e.g., [9, 14]. Let $\dot{\mathbf{x}} = d\mathbf{x}/dt$ denote the time derivative of a vector, while the Euclidean length is defined as $\|\mathbf{x}\| = \sqrt{\mathbf{x}^{\mathrm{T}}\mathbf{x}}$. Superscript denotes frame of reference for a given vector. The rotation matrix is denoted $\mathbf{R}^c_a \in \mathcal{SO}(3) = \left\{\mathbf{R}\in\mathbb{R}^{3\times3} : \mathbf{R}^{\mathrm{T}}\mathbf{R} = \mathbf{I}, \det(\mathbf{R}) = 1\right\}$, which rotates a vector from frame a to frame c and where $\mathbf{I}$ denotes the identity matrix. The inverse rotation is found by taking its transpose, such that $\mathbf{R}^a_c = \left(\mathbf{R}^c_a\right)^{\mathrm{T}}$. The angular velocity of frame c relative to frame a referenced in frame e is denoted $\boldsymbol{\omega}^e_{a,c}$, and angular velocities can be added together as $\boldsymbol{\omega}^e_{a,f} = \boldsymbol{\omega}^e_{a,c} + \boldsymbol{\omega}^e_{c,f}$ (cf. [15]).

The derivative of the rotation matrix is defined as $\dot{\mathbf{R}}_a^c = \mathbf{R}_a^c \mathbf{S}(\boldsymbol{\omega}_{c,a}^a)$ where $\mathbf{S}(\cdot)$ denotes the cross product operator, which is defined such that for two vectors $\mathbf{v}_1, \mathbf{v}_2 \in \mathbb{R}^3$, $\mathbf{S}(\mathbf{v}_1)\mathbf{v}_2 = \mathbf{v}_1 \times \mathbf{v}_2$, $\mathbf{S}(\mathbf{v}_1)\mathbf{v}_2 = -\mathbf{S}(\mathbf{v}_2)\mathbf{v}_1$, $\mathbf{S}(\mathbf{v}_1)\mathbf{v}_1 = \mathbf{0}$, and $\mathbf{v}_1^{\mathrm{T}}\mathbf{S}(\mathbf{v}_2)\mathbf{v}_1 = 0$.

Quaternions can be used to parameterize the rotation matrix, where $\mathbf{q}_{c,a} \in \mathcal{S}^3 = \{\mathbf{q} \in \mathbb{R}^4 : \mathbf{q}^{\mathrm{T}}\mathbf{q} = 1\}$ denotes the quaternion representing a rotation from frame a to frame c through the angle of rotation $\vartheta_{c,a}$ around the axis of rotation $\mathbf{k}_{c,a}$. The inverse quaternion is defined as $\mathbf{q}_{a,c} = \begin{bmatrix} \eta_{c,a} & -\boldsymbol{\varepsilon}_{c,a}^{\mathrm{T}} \end{bmatrix}^{\mathrm{T}}$, also sometimes denoted as $\mathbf{q}^*$. A quaternion comprises a scalar and a vector part, where $\eta_{c,a}$ denotes the scalar part, while $\boldsymbol{\varepsilon}_{c,a} \in \mathbb{R}^3$ denotes the vector part. This allows the rotation matrix to be constructed using quaternions as $\mathbf{R}_a^c = \mathbf{I} + 2\eta_{c,a}\mathbf{S}(\boldsymbol{\varepsilon}_{c,a}) + 2\mathbf{S}^2(\boldsymbol{\varepsilon}_{c,a})$. Composite quaternions can be found through quaternion multiplication as $\mathbf{q}_{c,e} = \mathbf{q}_{c,a} \otimes \mathbf{q}_{a,e} = \mathbf{T}(\mathbf{q}_{c,a})\mathbf{q}_{a,e}$ with

$$\mathbf{T}(\mathbf{q}_{c,a}) = \begin{bmatrix} \eta_{c,a} & -\boldsymbol{\varepsilon}_{c,a}^{\mathrm{T}} \\ \boldsymbol{\varepsilon}_{c,a} & \eta_{c,a}\mathbf{I} + \mathbf{S}(\boldsymbol{\varepsilon}_{c,a}) \end{bmatrix}. \tag{1}$$

The use of the quaternion product ensures that the resulting quaternion maintains the unit length property. The quaternion kinematics is defined as

$$\dot{\mathbf{q}}_{c,a} = \frac{1}{2}\mathbf{q}_{c,a} \otimes \begin{bmatrix} 0 \\ \boldsymbol{\omega}_{c,a}^a \end{bmatrix} = \frac{1}{2}\mathbf{T}(\mathbf{q}_{c,a}) \begin{bmatrix} 0 \\ \boldsymbol{\omega}_{c,a}^a \end{bmatrix}. \tag{2}$$

For attitude control, several different frames are needed:

Inertial: The Earth-centered inertial (ECI) has its origin in the center of the Earth, where the $\mathbf{x}^i$ axis points toward the vernal equinox and the $\mathbf{z}^i$ points through the North Pole, while $\mathbf{y}^i$ completes the right-handed orthonormal frame. The inertial frame is denoted by $\mathcal{F}^i$.

Orbit: The orbit frame has its origin in the center of mass of the satellite (cf. [16], p. 479). The $\mathbf{e}_r$ axis coincides with the radius vector $\mathbf{r}^i \in \mathbb{R}^3$, which goes from the center of the Earth to the center of mass in the satellite. The $\mathbf{e}_h$ axis is parallel to the orbital angular momentum vector, pointing in the normal direction of the orbit. The $\mathbf{e}_\theta$ completes the right-handed orthonormal frame where the vectors can be described as $\mathbf{e}_r = \frac{\mathbf{r}^i}{\|\mathbf{r}^i\|}$, $\mathbf{e}_\theta = \mathbf{e}_h \times \mathbf{e}_r$, and $\mathbf{e}_h = \frac{\mathbf{h}}{\|\mathbf{h}\|}$ where $\mathbf{h} = \mathbf{r}^i \times \dot{\mathbf{r}}^i$. The orbit frame is denoted by $\mathcal{F}^o$.

Body: The body frame has its origin in the center of mass of the satellite, where its axes coincide with the principal axes of inertia. The body frame is denoted by $\mathcal{F}^b$.

Desired: The desired frame can be defined arbitrarily to achieve any given objective (cf. [17]). The desired frame is denoted by $\mathcal{F}^d$.

2.2 Orbital mechanics

This section describes how the orbit frame can be related to the inertial frame through the six classical orbital parameters, and for more details, the reader is referred to [1]. Specifically, the objective with this subsection is to find the radius, velocity, and acceleration vector of the orbit, as well as its angular velocity and acceleration. From well-known orbital mechanics, the six classical parameters can be defined as the semimajor axis a, the eccentricity e, the inclination i, the right ascension of the ascending node Ω, the argument of the perigee ω, and the mean anomaly M. The distance to the apogee and perigee from the center of the Earth can be defined, respectively, as r_a and r_p, allowing the semimajor axis to be found as $a = \frac{r_a + r_p}{2}$ and the eccentricity of the orbit as $e = \frac{r_a - r_p}{r_a + r_p}$, while the mean motion can be found from $n = \sqrt{\frac{\mu}{a^3}}$. Here, $\mu = GM_{Earth}$, where G is the gravitational constant, while M_{Earth} is the mass of the Earth. With knowledge of the mean motion, the mean anomaly can be found as $M = n(t - t_0) = \psi - e \sin(\psi)$ where ψ is the eccentric anomaly, t is the time, and t_0 is the time when passing the perigee. To properly describe where in the orbit the satellite is located, the true anomaly can be found as $\theta = \cos^{-1}\left(\frac{\cos(\psi) - e}{1 - e\cos(\psi)}\right)$, while its derivative can be found as $\dot{\theta} = \frac{n(1 + e\cos(\theta))^2}{(1 - e^2)^{\frac{3}{2}}}$ ([18], p. 42). When running a simulation, it is desirable to have a continuously increasing true anomaly, while the direct method will map the angle between 0 and 180°. Instead, by integrating the derivative overtime, a smooth true anomaly can be found that increases continuously. The eccentric anomaly, however, cannot be found analytically, but can be found through an iterative algorithm as described in ([1], p. 26) $\psi_k(t) = M(t) + e\sin(\psi_{k-1}(t))$, where k is the iteration number. This algorithm is valid as long as $0 < e < 1$, which holds for elliptical orbits. From these calculations, the rotation matrix from inertial frame to orbit frame can now be constructed as

$$\mathbf{R}_i^o = \begin{bmatrix} \cos(\omega+\theta)\cos(\Omega) - \cos(i)\sin(\omega+\theta)\sin(\Omega) & \cos(\omega+\theta)\sin(\Omega) + \sin(\omega+\theta)\cos(i)\cos(\Omega) & \sin(\omega+\theta)\sin(i) \\ -\sin(\omega+\theta)\cos(\Omega) - \cos(i)\sin(\Omega)\cos(\omega+\theta) & -\sin(\omega+\theta)\sin(\Omega) + \cos(\omega+\theta)\cos(i)\cos(\Omega) & \cos(\omega+\theta)\sin(i) \\ \sin(i)\sin(\Omega) & -\sin(i)\cos(\Omega) & \cos(i) \end{bmatrix}. \tag{3}$$

The radius, velocity, and acceleration vector can be defined in the orbit frame, respectively, as ([1], pp. 26–27)

$$\mathbf{r}^o = \begin{bmatrix} a\cos(\psi) - ae & a\sin(\psi)\sqrt{1-e^2} & 0 \end{bmatrix}^{\mathrm{T}}, \tag{4}$$

$$\mathbf{v}^o = \begin{bmatrix} -\frac{a^2 n}{r}\sin(\psi) & \frac{a^2 n}{r}\sqrt{1-e^2}\cos(\psi) & 0 \end{bmatrix}^{\mathrm{T}}, \tag{5}$$

$$\mathbf{a}^o = \begin{bmatrix} -\frac{a^3 n^2}{r^2}\cos(\psi) & -\frac{a^3 n^2}{r^2}\sqrt{1-e^2}\sin(\psi) & 0 \end{bmatrix}^{\mathrm{T}}, \tag{6}$$

where $r = \|\mathbf{r}^i\|$ is the length of the radius vector. Each of these vectors can be rotated to the inertial frame using Eq. (3), such that $\mathbf{r}^i = \mathbf{R}_i^o \mathbf{r}^o$, $\mathbf{v}^i = \mathbf{R}_o^i \mathbf{v}^o$, and $\mathbf{a}^i = \mathbf{R}_o^i \mathbf{a}^o$. The angular velocity of the orbit frame relative to the inertial frame can be found as $\boldsymbol{\omega}_{i,o}^i = \frac{\mathbf{r}^i \times \mathbf{v}^i}{(\mathbf{r}^i)^{\mathrm{T}}\mathbf{r}^i}$, while the angular acceleration can be found through differentiation as

$$\dot{\boldsymbol{\omega}}_{i,o}^i = \frac{(\mathbf{r}^i \times \mathbf{a}^i)(\mathbf{r}^i)^{\mathrm{T}}\mathbf{r}^i - 2(\mathbf{r}^i \times \mathbf{v}^i)(\mathbf{v}^i)^{\mathrm{T}}\mathbf{r}^i}{\left((\mathbf{r}^i)^{\mathrm{T}}\mathbf{r}^i\right)^2}. \tag{7}$$

In order to implement the orbital mechanics in, e.g., a Simulink framework, the required input to the subsystem would be the time (t). Further, the orbital parameters must be defined as given in **Table 1** and can be changed depending on the orbit. These constants allow for the calculations of the eccentricity (e), the semimajor axis (a), and mean motion (n). With the mean motion, the mean anomaly (M) can be found and used to approximate the eccentric anomaly (ψ) using the iterative algorithm presented above. The rate of change of the true anomaly ($\dot{\theta}$) can also be found and by integration enables the calculation of true anomaly (θ). All

Parameter	Value	Unit
G	$6.67408 \cdot 10^{-11}$	$m^3\ kg^{-1}\ s^{-2}$
M_{Earth}	$5.9742 \cdot 10^{24}$	kg
r_a	$R_e + 1200$	km
r_p	$R_e + 800$	km
R_e	6378	km
i	75	°
Ω	0	°
ω	0	°

Table 1.
Parameters required for calculation of the orbital dynamics.

these values allow for calculating the rotation matrix from the orbit frame to the inertial frame ($\mathbf{R}_o^i$), the radius vector ($\mathbf{r}^o$), the velocity vector ($\mathbf{v}^o$), the acceleration vector ($\mathbf{a}^o$), angular velocity ($\boldsymbol{\omega}_{i,o}^i$), and angular acceleration vector ($\dot{\boldsymbol{\omega}}_{i,o}^i$). Hence, all the outputs from this subsystem can easily be found following this procedure and will be used in several other subsystems.

3. Attitude dynamics and control

3.1 Attitude dynamics

The attitude dynamics can be derived with the basis in Euler's moment Equation ([1], p. 95). The angular momentum of a rigid body in the body frame is given as

$$\mathbf{h}^b = \mathbf{J}\boldsymbol{\omega}_{i,b}^b, \tag{8}$$

where $\mathbf{J}\in\mathbb{R}^{3x3}$ is the inertia matrix, while $\boldsymbol{\omega}_{i,b}^b\in\mathbb{R}^3$ is the angular velocity of the body frame relative to the inertial frame. The angular momentum can be found in the inertial frame as

$$\mathbf{h}^i = \mathbf{R}_b^i\mathbf{h}^b. \tag{9}$$

The rate of change of angular momentum is equal to the total torque, such that $\boldsymbol{\tau}^i = \dot{\mathbf{h}}^i$. Hence, by differentiating Eq. (9), it is obtained that

$$\boldsymbol{\tau}^i = \dot{\mathbf{h}}^i = \mathbf{R}_b^i\mathbf{S}(\boldsymbol{\omega}_{i,b}^b)\mathbf{h}^b + \mathbf{R}_b^i\dot{\mathbf{h}}^b, \tag{10}$$

which can be written in the body frame by using Eq. (8) as $\boldsymbol{\tau}^b = \mathbf{S}\left(\boldsymbol{\omega}_{i,b}^b\right)\mathbf{J}\boldsymbol{\omega}_{i,b}^b + \mathbf{J}\dot{\boldsymbol{\omega}}_{i,b}^b$, where the inertia matrix is assumed to be constant. Decomposing the total torque into an actuation component and a perturbing component, $\boldsymbol{\tau}^b = \boldsymbol{\tau}_a^b + \boldsymbol{\tau}_p^b$, allows the rotational dynamics to be written as

$$\mathbf{J}\dot{\boldsymbol{\omega}}_{i,b}^b = -\mathbf{S}(\boldsymbol{\omega}_{i,b}^b)\mathbf{J}\boldsymbol{\omega}_{i,b}^b + \boldsymbol{\tau}_a^b + \boldsymbol{\tau}_p^b, \tag{11}$$

where $\boldsymbol{\tau}_a^b \in \mathbb{R}^3$ denotes the actuation torques (e.g., output from reaction wheels), while $\boldsymbol{\tau}_p^b \in \mathbb{R}^3$ denotes the perturbing torques (e.g., gravity torque). Further, by using quaternion representation, the update law for the quaternion representing the attitude of the body frame relative to the inertial frame can be written as

$$\dot{\mathbf{q}}_{i,b} = \frac{1}{2}\mathbf{T}\left(\mathbf{q}_{i,b}\right)\begin{bmatrix} 0 \\ \boldsymbol{\omega}_{i,b}^b \end{bmatrix}. \tag{12}$$

Hence, Eqs. (11) and (12) serve as governing equations describing the attitude and angular velocity of the satellite. The inputs that affect these values are the perturbation and actuation torques, where the latter will be found in the following sections.

3.2 Error dynamics

From Euler's moment equation, the angular acceleration is defined relative to the inertial frame. For attitude control, it is often more interesting controlling the attitude and angular velocity relative to the orbit frame. For that reason, the angular velocity of the body frame relative to the orbit frame can be found as $\boldsymbol{\omega}_{o,b}^b = \boldsymbol{\omega}_{i,b}^b - \mathbf{R}_i^b \boldsymbol{\omega}_{i,o}^i$, which can be differentiated as

$$\mathbf{J}\dot{\boldsymbol{\omega}}_{o,b}^b = -\mathbf{S}(\boldsymbol{\omega}_{i,b}^b)\mathbf{J}\boldsymbol{\omega}_{i,b}^b + \boldsymbol{\tau}_a^b + \boldsymbol{\tau}_p^b + \mathbf{J}\mathbf{S}(\boldsymbol{\omega}_{i,b}^b)\mathbf{R}_i^b \boldsymbol{\omega}_{o,i}^i - \mathbf{J}\mathbf{R}_i^b \dot{\boldsymbol{\omega}}_{i,o}^i \tag{13}$$

giving a description of the attitude dynamics relative to the orbit frame. It is also possible to find the error dynamics, to enable tracking of a desired attitude and angular velocity. Let $\mathbf{q}_{o,d}, \boldsymbol{\omega}_{o,d}^d, \dot{\boldsymbol{\omega}}_{o,d}^d \in \mathcal{L}_\infty$ denote a desired quaternion, angular velocity, and acceleration; then, the quaternion and angular velocity error can be found as

$$\mathbf{q}_{d,b} = \mathbf{q}_{d,o} \otimes \mathbf{q}_{o,b}, \tag{14}$$

$$\boldsymbol{\omega}_{d,b}^b = \boldsymbol{\omega}_{o,b}^b - \mathbf{R}_d^b \boldsymbol{\omega}_{o,d}^d, \tag{15}$$

with the kinematics as

$$\dot{\eta}_{d,b} = -\frac{1}{2}\boldsymbol{\varepsilon}_{d,b}^{\mathrm{T}} \boldsymbol{\omega}_{d,b}^b, \tag{16}$$

$$\dot{\boldsymbol{\varepsilon}}_{d,b} = \left(\eta_{d,b}\mathbf{I} + \mathbf{S}(\boldsymbol{\varepsilon}_{d,b})\right)\boldsymbol{\omega}_{d,b}^b. \tag{17}$$

The angular acceleration error can be found by differentiating Eq. (15) as

$$\mathbf{J}\dot{\boldsymbol{\omega}}_{d,b}^b = -\mathbf{S}(\boldsymbol{\omega}_{i,b}^b)\mathbf{J}\boldsymbol{\omega}_{i,b}^b + \boldsymbol{\tau}_a^b + \boldsymbol{\tau}_p^b + \mathbf{J}\mathbf{S}(\boldsymbol{\omega}_{i,b}^b)\mathbf{R}_i^b \boldsymbol{\omega}_{o,i}^i - \mathbf{J}\mathbf{R}_i^b \dot{\boldsymbol{\omega}}_{i,o}^i + \mathbf{J}\mathbf{S}\left(\boldsymbol{\omega}_{o,b}^b\right)\mathbf{R}_d^b \boldsymbol{\omega}_{o,d}^d - \mathbf{J}\mathbf{R}_d^b \dot{\boldsymbol{\omega}}_{o,d}^d. \tag{18}$$

Hence, the control objective can be defined as that of making $\left(\mathbf{q}_{d,b}, \boldsymbol{\omega}_{d,b}^b\right) \to (\mathbf{0}, \mathbf{0})$, which will make the satellite point in a desired direction and move with a desired angular velocity.

3.3 PD+ attitude controller

Takegaki and Arimoto [19] proposed in 1981 a simple method for position control of robots, something that was extended by [20] to enable trajectory tracking. The so-called PD+ controller has been applied for spacecraft by [21, 22] showing good results. The author has applied this method in previous works such as [23, 24].

In order to design a PD+ attitude controller, let a Lyapunov function candidate be chosen as $V = \frac{1}{2}\left(\boldsymbol{\omega}^b_{d,b}\right)^{\mathrm{T}}\mathbf{J}\boldsymbol{\omega}^b_{d,b} + k_p\left(1-\eta_{d,b}\right)^2 + k_p\boldsymbol{\varepsilon}^{\mathrm{T}}_{d,b}\boldsymbol{\varepsilon}_{d,b}$ where k_p is a positive scalar gain. The derivative is found by using Eqs. (16)–(18) as

$$\begin{aligned}\dot{V} &= k_p\boldsymbol{\varepsilon}^{\mathrm{T}}_{d,b}\boldsymbol{\omega}^b_{d,b} + \left(\boldsymbol{\omega}^b_{d,b}\right)^{\mathrm{T}}\Big(-\mathbf{S}(\boldsymbol{\omega}^b_{i,b})\mathbf{J}\boldsymbol{\omega}^b_{i,b} + \boldsymbol{\tau}^b_a + \boldsymbol{\tau}^b_p + \mathbf{J}\mathbf{S}(\boldsymbol{\omega}^b_{i,b})\mathbf{R}^b_i\boldsymbol{\omega}^i_{o,i} \\ &\quad -\mathbf{J}\mathbf{R}^b_i\dot{\boldsymbol{\omega}}^i_{i,o} + \mathbf{J}\mathbf{S}(\boldsymbol{\omega}^b_{o,b})\mathbf{R}^b_d\boldsymbol{\omega}^d_{o,d} - \mathbf{J}\mathbf{R}^b_d\dot{\boldsymbol{\omega}}^d_{o,d}\Big).\end{aligned} \tag{19}$$

A PD+ attitude control law can now be chosen as

$$\begin{aligned}\boldsymbol{\tau}^b_d &= \mathbf{J}\mathbf{R}^b_d\dot{\boldsymbol{\omega}}^d_{o,d} - \mathbf{J}\mathbf{S}(\boldsymbol{\omega}^b_{o,b})\mathbf{R}^b_d\boldsymbol{\omega}^d_{o,d} + \mathbf{J}\mathbf{R}^b_i\dot{\boldsymbol{\omega}}^i_{i,o} - \mathbf{J}\mathbf{S}(\boldsymbol{\omega}^b_{i,b})\mathbf{R}^b_i\boldsymbol{\omega}^i_{o,i} - \boldsymbol{\tau}^b_p + \mathbf{S}(\boldsymbol{\omega}^b_{i,b})\mathbf{J}\boldsymbol{\omega}^b_{i,b} \\ &\quad -k_p\boldsymbol{\varepsilon}_{d,b} - k_d\boldsymbol{\omega}^b_{d,b},\end{aligned} \tag{20}$$

where k_d is another positive scalar gain and $\boldsymbol{\tau}^b_d$ denotes the desired torque required to make the attitude and angular velocity errors go to zero. Assuming no actuator dynamics, i.e., $\boldsymbol{\tau}^b_a = \boldsymbol{\tau}^b_d$, and then by inserting Eq. (20) into Eq. (19), it is obtained that $\dot{V} \leq -k_d\left\|\boldsymbol{\omega}^b_{d,b}\right\|^2$, which is negative semidefinite. By applying the Matrosov theorem (cf. [24]), it can be shown that the origin $(\boldsymbol{\varepsilon}_{d,b}, \boldsymbol{\omega}^b_{d,b}) = (\mathbf{0}, \mathbf{0})$ is uniformly asymptotically stable.

The inputs to the control law (Eq. 20) are the desired states $\mathbf{q}_{o,d}$, $\boldsymbol{\omega}^d_{o,d}$, and $\dot{\boldsymbol{\omega}}^d_{o,d}$, which are to be defined by the reader, e.g., as part as a guidance block depending on the mission objective. The inertia matrix ($\mathbf{J}$) is assumed to be known, while the angular velocity vector between the body frame and orbit frame ($\boldsymbol{\omega}^b_{o,b}$) can be found as described above. The other angular velocities are found from the orbital mechanics, while the rotation matrices are found as composite rotations, e.g., $\mathbf{R}^b_i = \mathbf{R}^b_o\mathbf{R}^o_i$, or by using the relationship between the quaternions and rotation matrices directly (cf. Section 2A). The error quaternion and angular velocity are found from Eqs. (14) and (15), while the perturbing torques will be described in the following section.

4. Perturbing torques

There are different kinds of perturbing torques, such as gravity torque, aerodynamic torque, magnetic field due to the electronics inside the satellite, as well as solar radiation torque. This section only considers the gravity torque. In [16], p. 147, the gravity torque is defined as

$$\boldsymbol{\tau}^b_g = \frac{3GM_{Earth}}{r^5}\mathbf{r}^i \times \mathbf{J}\mathbf{r}^i, \tag{21}$$

where the terms have previously been defined. As can be seen from this equation, non-diagonal inertia matrices will induce gravitational torques to align the satellite with the gravity field. This is also sometimes used for passive control, using e.g., a gravity boom to ensure that one side of the satellite is always facing the Earth. For this chapter, the perturbing toque is set equal to the gravity torque such that $\tau_p^b = \tau_g^b$.

5. Actuators

The control signal must be mapped to an actuator that must generate the desired torque. With limitations in actuation, the saturation must be modeled in order to obtain realistic results when simulating attitude control. This section considers three types of actuators commonly used for spacecraft attitude control: magnetic torquers, reaction wheels, and thrusters.

5.1 Magnetic torquers

Magnetic torquers operate by creating a local magnetic field that interacts with the Earth's magnetic field. In simple terms, magnetic torques can be explained as that of a compass needle. By applying current through a coil, a local magnetic field is created, which will try to align itself with the Earth's magnetic field. This allows the attitude of a spacecraft to be changed and is a very popular approach for small satellites, e.g., cubesats. One of the drawbacks or challenges with magnetic actuation lies in the fact that the Earth's magnetic field goes from the North Pole to the South Pole as shown in **Figure 2.**[1] As can be seen, when the satellite crosses the North Pole, there will be mainly a downward magnetic field component, reducing the possibility of actuation to only two axes, and similarly along the equator. This subsection is based on [12] and will describe how to model magnetic torquers and how it can be applied for attitude control. A magnetic torquer produces a magnetic torque by applying a current through a coil, which can be expressed as [2].

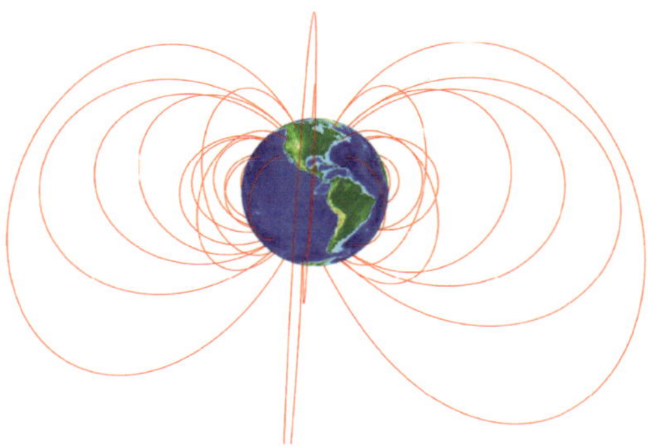

Figure 2.
Magnetic field of the Earth visualized using the IGRF model. The control torque using magnetic torquers is always perpendicular to the magnetic field, such that a the poles, only roll, and pitch control are available, while at the equator, only pitch and yaw control is available [25].

[1] Figure created using the MATLAB script "international geomagnetic reference field (IGRF) model" by Drew Compston.

$$\boldsymbol{\tau}_m^b = \mathbf{S}(\mathbf{m}^b)\mathbf{b}^b, \tag{22}$$

where $\mathbf{m}^b = \begin{bmatrix} m_x & m_y & m_z \end{bmatrix}^{\mathrm{T}} = NA\begin{bmatrix} i_x & i_y & i_z \end{bmatrix}^{\mathrm{T}}$ is the induced magnetic field, N is the number of turns of the coils, $i_{(\cdot)}$ is the current around a given axis, and A is the area of the coils. The Earth's magnetic field is represented through the vector $\mathbf{b}^b$, meaning that to use magnetic actuation for attitude control, a good model of the Earth's magnetic field is needed.

From a control point of view, the physical parameters A and N are defined when the spacecraft is designed, such that the controller needs to dictate which currents that must be sent to the torquers in order to get a desired torque. This means that Eq. (22) must be inverted with regard to $\mathbf{m}^b$, which is not straightforward due to the cross product, meaning that you obtain rank 2 when inverting the right-hand side, losing information about one of the axes. To that end, an approximation to inverting this equation is given in [25].

$$\mathbf{m}^b = \frac{\mathbf{S}\left(\mathbf{b}^b\right)\boldsymbol{\tau}_d^b}{\left\|\mathbf{b}^b\right\|^2}, \tag{23}$$

enabling the currents to be found as

$$\begin{bmatrix} i_x & i_y & i_z \end{bmatrix}^{\mathrm{T}} = \frac{1}{NA}\begin{bmatrix} m_x & m_y & m_z \end{bmatrix}^{\mathrm{T}}. \tag{24}$$

It is here assumed that all three torquers are identical, but depending on satellite configuration, there might be differences in the number of turns and areas. Hence, the desired torque $\boldsymbol{\tau}_d^b$ can be used to find the magnetic moment $\mathbf{m}^b$ in Eq. (23) and solved for the currents and applied resulting in the actuation torque in Eq. (22). Hence, the control law (Eq. 20) can be mapped to a desired magnetic moment (Eq. 23), which then can be used to find the desired current to each of the three coils. Then, the limits in current will dictate the maximum magnetic moment that can be generated.

Consider the HiNCube satellite as shown in **Figure 3**. The cubesat comprises three orthonormal magnetic torquers with an area $A = 0.00757$ m^2 and with a maximum current of 47.27 mA and $N = 100$ turns. This gives a maximum magnetic moment of $m_{max} = 0.03578$ mA^2. Hence, an implementation of using magnetic torquers for attitude control would encompass mapping the output from the control law to a desired magnetic moment using Eq. (23) and then imposing the maximum magnetic moment on each axis, before finding the resulting actuation torque using Eq. (22). Note that to ensure sign correctness due to the projection, the actuation torque can be found as

$$\begin{bmatrix} \tau_{x,a} & \tau_{y,a} & \tau_{z,a} \end{bmatrix}^{\mathrm{T}} = \begin{bmatrix} \mathrm{sign}(\tau_{x,d})\tau_{x,m} & \mathrm{sign}(\tau_{y,d})\tau_{y,m} & \mathrm{sign}(\tau_{z,d})\tau_{z,m} \end{bmatrix}^{\mathrm{T}}. \tag{25}$$

where $\boldsymbol{\tau}_a^b = \begin{bmatrix} \tau_{x,a} & \tau_{y,a} & \tau_{z,a} \end{bmatrix}^{\mathrm{T}}$, $\boldsymbol{\tau}_d^b = \begin{bmatrix} \tau_{x,d} & \tau_{y,d} & \tau_{z,d} \end{bmatrix}^{\mathrm{T}}$, and $\boldsymbol{\tau}_m^b = \begin{bmatrix} \tau_{x,m} & \tau_{y,m} & \tau_{z,m} \end{bmatrix}^{\mathrm{T}}$.

To show the performance of magnetic torquers, consider again the HiNCube satellite, which had an inertia matrix of $\mathbf{J} = 1.67 \cdot 10^{-3}\mathbf{I}$ kg m^2. Consider the problem of making rotating 90° from an initial quaternion $\mathbf{q}_{o,b} = [0 \quad 0 \quad 0 \quad 1]^{\mathrm{T}}$ to $\mathbf{q}_{o,d} = [1 \quad 0 \quad 0 \quad 0]^{\mathrm{T}}$. The gains for the PD+ controller are set $k_p = 1 \cdot 10^{-5}$ and

$k_d = 5 \cdot 10^{-3}$, and the satellite is assumed to have an orbit of $r_p = 500$ km and $r_a = 600$ km, with inclination of 75°. **Figure 4** shows the attitude, angular velocity, and actuation torque. It is evident that magnetic torquers produce very low torque, such that it takes a very long time to change the attitude of the spacecraft (about 1 h). To some extents, this can be improved by being in a lower orbit where the magnetic

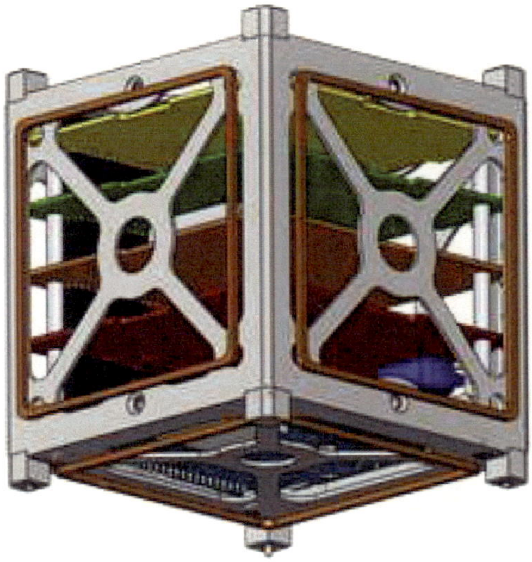

Figure 3.
Magnetic torquers on the HiNCube satellite (shown in brown).

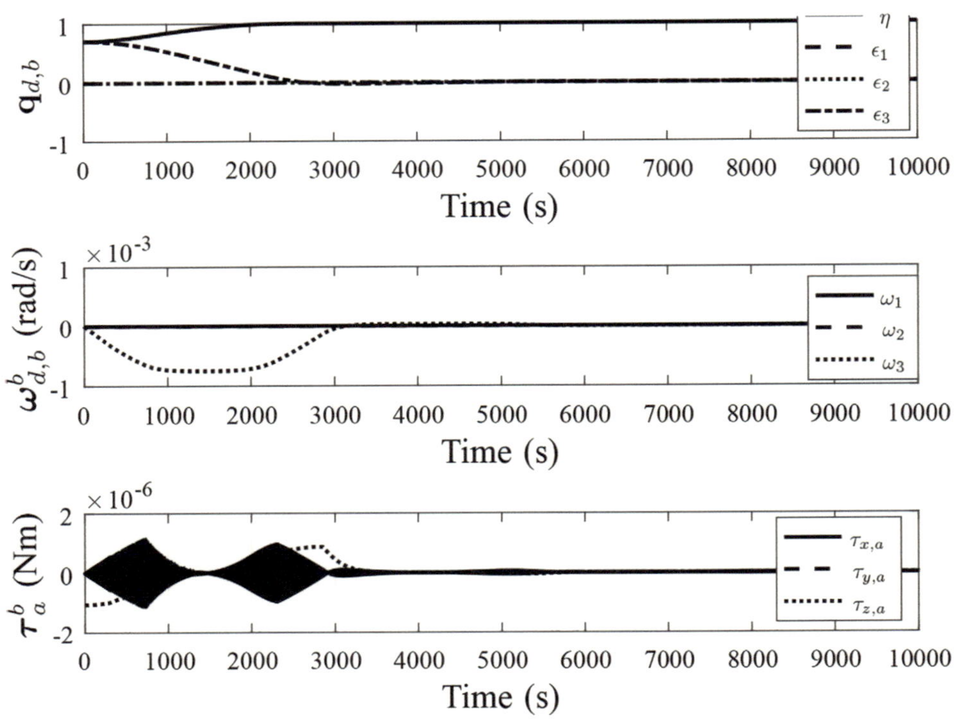

Figure 4.
Quaternion error, angular velocity error, and actuation torque using magnetic torquers.

field is stronger or by using larger coils with higher currents. Also, note that the actuation signal varies in strength as a function of time, depending on the orbital position of the satellite.

5.2 Reaction wheels

Another way of changing the attitude of a satellite is through reaction wheels. Reaction wheels are based on the principle of Newton's third law: *When one body exerts a force on a second body, the second body simultaneously exerts a force equal in magnitude and opposite in direction on the first body*. This means that by spinning a reaction wheel in one direction, the satellite will rotate in the other direction. Mounting three reaction wheels in an orthogonal configuration enables three-axis attitude control of spacecraft. From Newton's third law, the momentum generated by the reaction wheels will have opposite sign of the momentum of the satellite, such that $\dot{\mathbf{h}}^i = -\dot{\mathbf{h}}_w^i$ where $\dot{\mathbf{h}}_w^i$ is the momentum production by the reaction wheels, $\dot{\mathbf{h}}^i$ is the momentum acting on the satellite, and $\boldsymbol{\tau}_w^b$. By employing Euler's moment equation similarly as in Section 3, the torque generated by the reaction wheels can be found by differentiating $\mathbf{h}_w^i = \mathbf{R}_b^i\mathbf{h}_w^b$ with $\mathbf{h}_w^b = \mathbf{J}_w\boldsymbol{w}_w^b$ where $\boldsymbol{\omega}_w^b$ is the angular velocity of the reaction wheels and $\mathbf{J}_w$ denotes their inertia. This gives

$$\boldsymbol{\tau}_a^b = -\dot{\mathbf{h}}_w^b - \mathbf{S}(\boldsymbol{\omega}_{i,b}^b)\mathbf{h}_w^b, \tag{26}$$

where $\boldsymbol{\tau}_w^b = \dot{\mathbf{h}}_w^b$ is the torque generated by the reaction wheels. Now, consider a set of three orthonormal reaction wheels, where one produces torques around the x-axis, one around the y-axis, and one around the z-axis of the body frame. Then, the PD+ control law dictates a desired torque, $\boldsymbol{\tau}_d^b$, which shall be achieved by the reaction wheels. To that end, the torque by the reaction wheel can be rewritten as $\boldsymbol{\tau}_w^b = \boldsymbol{\tau}_d^b - \mathbf{S}\left(\boldsymbol{\omega}_{i,b}^b\right)\mathbf{h}_w^b$, where $\boldsymbol{\tau}_w^b$ must be bounded by the motor torque limit, while the angular momentum will be bounded as a function of maximum rotational speed. After imposing the torque and speed constraints, the angular momentum of the reaction wheels is found by integrating $\dot{\mathbf{h}}_w^b$ allowing the actuation torque to be calculated using Eq. (26).

Consider the HiNCube satellite again, where it is possible to use three small reaction wheels as described in [11] where the main idea is to place most of the mass away from the center as shown in **Figure 5**. The inertia of an individual reaction wheel was found to be $J_w = 1.46 \cdot 10^{-5}$ kg m^2, and by assuming a maximum rotation speed of 13,700 rpm with maximum torque of $\tau_{max} = 0.0047$ Nm, the maximum momentum generated by the reaction wheels is found as $h_{max} = J_w\omega_w = 1.46 \cdot 10^{-5} \cdot 13700 \cdot \frac{2\pi}{60} = 1.52389 \cdot 10^{-6}$. Now, consider the same simulation as when using the magnetic torquers, where the gains for the PD+ controller is changed to $k_p = k_d = 2$ and the reaction wheels has the limits as defined above. **Figure 6** shows the simulation results, where it is obvious that by using reaction wheels, the satellite is able to change its orientation after about 80 s. To some extent, this can be credited to the higher gains, but it lies mainly with the better actuation system that is able to produce higher torque than the reaction wheels. From the figure, the reaction wheels quickly go into saturation of 13, 700 RPM, where the angular velocity also goes into saturation. As the quaternion error goes toward zero, the reaction wheel despin, reducing the angular velocity and the control objective, is completed.

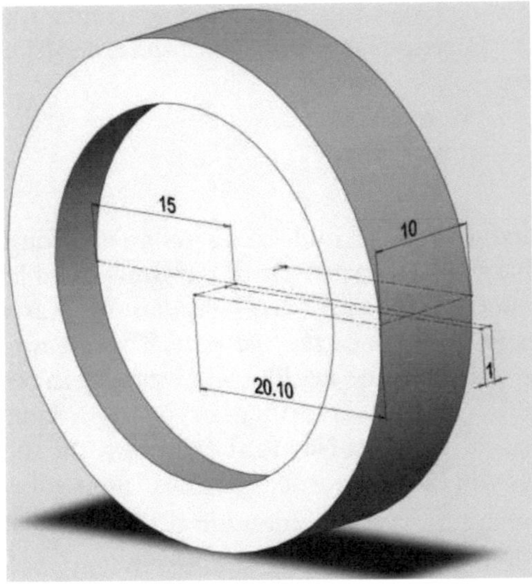

Figure 5.
Example design of a reaction wheel for cubesats (dimensions are in mm) [11].

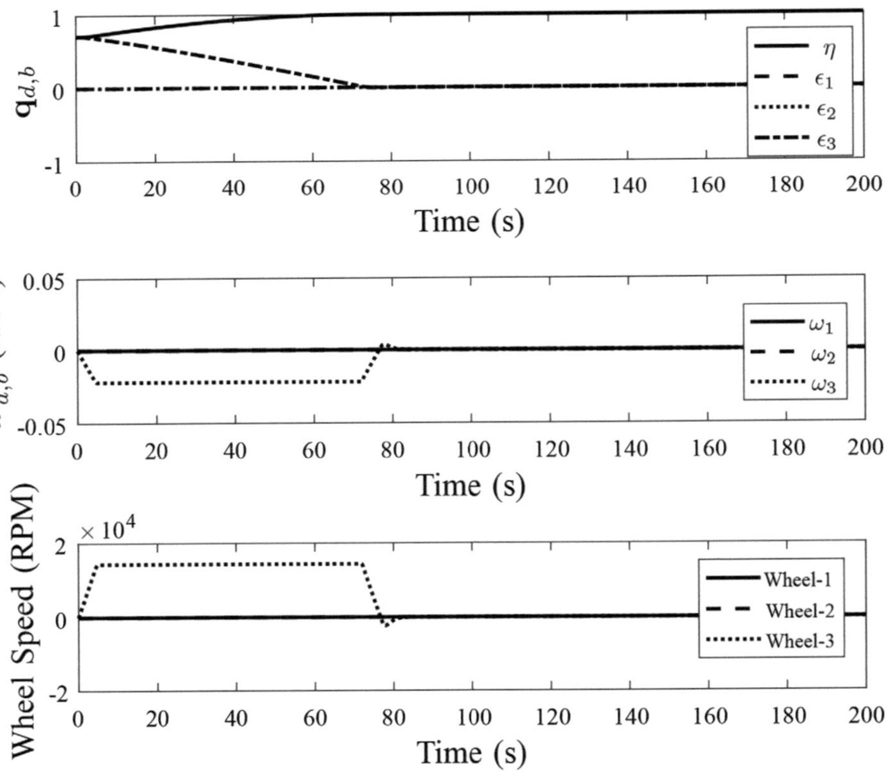

Figure 6.
Quaternion error, angular velocity error, and wheel speeds when using reaction wheels for attitude control.

5.3 Thrusters

The third kind of actuator that will be studied is using reaction control thrusters. This section presents how to map the control signal (Eq. 20) to four thrusters used for attitude control. Let the location of each thruster be denoted by

$\mathbf{r}_i^b = \begin{bmatrix} r_x & r_y & r_z \end{bmatrix}^{\mathrm{T}}$, and let them have an azimuth and an elevation angle described by χ and γ. Then, the torque produced by a given thruster can be found as [1], p. 262

$$\boldsymbol{\tau}_i^b = \mathbf{r}_i^b \times \mathbf{f}_i^b = \begin{bmatrix} r_y \sin(\chi)\cos(\gamma) - r_z \sin(\gamma) \\ r_z \cos(\gamma)\cos(\chi) - r_x \cos(\gamma)\sin(\chi) \\ r_x \sin(\gamma) - r_y \cos(\gamma)\cos(\chi) \end{bmatrix} f_i, \tag{27}$$

where f_i denotes the total thrust from the ith thruster. Given the thruster configuration defined in **Table 2**, let the vector of thruster signals be denoted $\mathbf{u} = \begin{bmatrix} f_1 & f_2 & f_3 & f_4 \end{bmatrix}^{\mathrm{T}}$, and then the torque can be found as $\boldsymbol{\tau}_a^b = \mathbf{B}\mathbf{u}$ with

$$\mathbf{B} = \begin{bmatrix} -\frac{\sqrt{2}}{5} & \frac{\sqrt{2}}{5} & \frac{\sqrt{2}}{5} & -\frac{\sqrt{2}}{5} \\ \frac{\sqrt{2}}{4} & -\frac{\sqrt{2}}{4} & \frac{\sqrt{2}}{4} & -\frac{\sqrt{2}}{4} \\ -\frac{\sqrt{2}}{4} & -\frac{\sqrt{2}}{4} & \frac{\sqrt{2}}{4} & \frac{\sqrt{2}}{4} \end{bmatrix}. \tag{28}$$

Given a desired torque from the PD+ control law, it must be mapped to the desired thruster firings, such that the combination of thrusters produces the desired torque. To that end, there are several different modulation methods that can be applied, ranging from a simple bang-bang modulation to more sophisticated pulse-width pulse-frequency modulation. This section will give an introduction to the different methods and detail how they can be implemented. In general the desired torque can be mapped to the desired thruster firings as $\mathbf{u}_d = \mathbf{B}^{\dagger}\boldsymbol{\tau}_d^b$ where $\dagger$ denotes the Moore-Penrose pseudoinverse and $\mathbf{u}_d = \begin{bmatrix} u_1 & u_2 & u_3 & u_4 \end{bmatrix}^{\mathrm{T}}$ denotes the magnitude of each of the thrusters.

1. *Bang-bang modulation*: The easiest approach to thruster firings is bang-bang modulation, where the thruster is fully actuated as long as the ith signal of $\mathbf{u}_d$ is above zero, such that

$$f_i = \begin{cases} f_{max} & \text{if} \quad u_i > 0 \\ 0 & \text{if} \quad u_i \leq 0 \end{cases}, \tag{29}$$

where f_{max} denotes the maximum available thrust from the ith thruster. After applying bang-bang modulation, the vector $\mathbf{u}$ can be constructed allowing the actuator torque to be found as $\boldsymbol{\tau}_a^b = \mathbf{B}\mathbf{u}$.

Thruster	Elevation (γ)	Azimuth (χ)	r_x	r_y	r_z
f_1	45	90	−0.5	−0.45	−0.05
f_2	135	90	−0.5	−0.45	0.05
f_3	−45	90	−0.5	0.45	−0.05
f_4	−135	90	−0.5	0.45	0.05

Table 2.
Thruster configuration.

2. *Bang-bang modulation with dead zone*: One of the major drawbacks by using simple bang-bang modulation is when the tracking error has converged to zero, where the thruster firings will continue to maintain the desired attitude. Sensor noise is another source that leads to continuous firings, quickly spending all the propellant. To that end, bang-bang modulation can be augmented with a dead zone, giving

$$f_i = \begin{cases} f_{max} & if \quad u_i > D \\ 0 & if \quad u_i \leq D \end{cases}, \tag{30}$$

where $D > 0$ denotes the dead zone. By properly selecting a suitable dead zone enables the thrusters to avoid firing when close to the equilibrium point.

3. *Pulse-width modulation*: Another approach that is often used for thruster firings is by using pulse-width modulation (PWM), where an analogue signal (desired torque) can be mapped to discrete signals using PWM. Instead of changing the thrust level, the duration of the pulses can be changed, leading to a pulse that is proportional to the torque command from the PD+ controller. A simple way of achieving this is by using the intersective technique, which uses a sawtooth signal that is compared to the control signal. When the sawtooth is less than the control signal, the PWM signal is in a high state and otherwise in a low state. This makes it possible to go from continuous control signal to a discrete representation which can be used for thruster firings. **Figure 7** shows how to achieve the PWM signal, enabled through a simple comparison of the two signals.

4. *Pulse-width pulse-frequency modulation*: In addition to controlling the width of the pulse as in PWM, it is also possible to control the frequency of the pulse—something that is done through pulse-width pulse-frequency (PWPF) modulation ([1], p. 265) (**Figure 8**). The modulation approach comprises a lag filter and a Schmitt trigger as shown in **Figure 9**. As long as the input to the Schmitt trigger is below U_{on}, the output is kept at zero and must be larger than $\frac{U_{on}}{K}$ to produce an output, where K is a DC gain, τ is the time constant, U_{on} and U_{off} are the on and off limits for the Schmitt trigger, while U_m is the maximum output. Much research has been performed on improving PWPF modulation, and in [26], the authors propose the following settings (cf. **Figure 9**): $2 < K < 6$, $0.1 < \tau < 0.5$, $U_{on} > 0.3$, $U_{off} < 0.8U_{on}$, and $U_m = 1$.

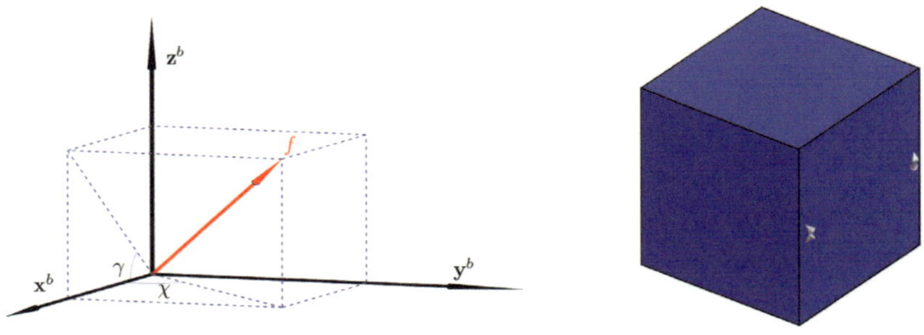

Figure 7.
Thruster configuration. The left subfigure shows the definition of azimuth and elevation angles used to dictate the orientation of the thruster, while the right subfigure shows a satellite with thrusters placed and oriented as given in ***Table 2****.*

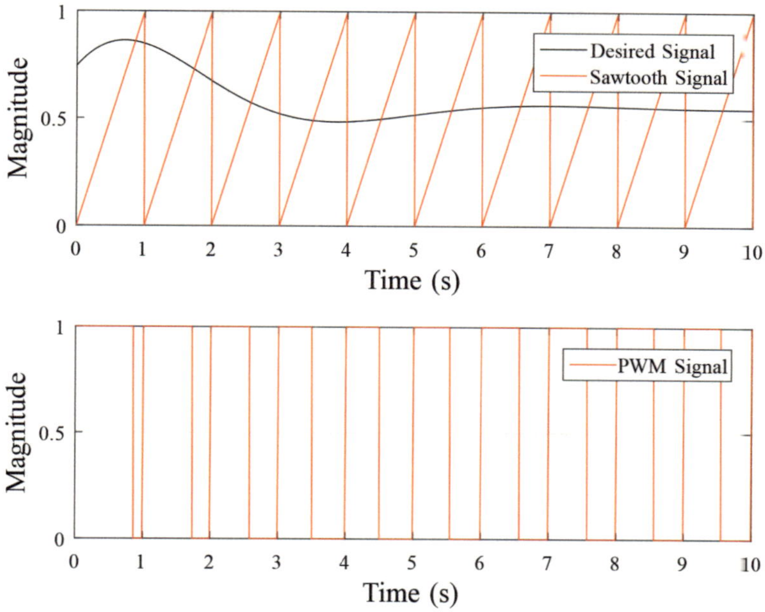

Figure 8.
Achieving pulse-width modulated signals for thruster firings.

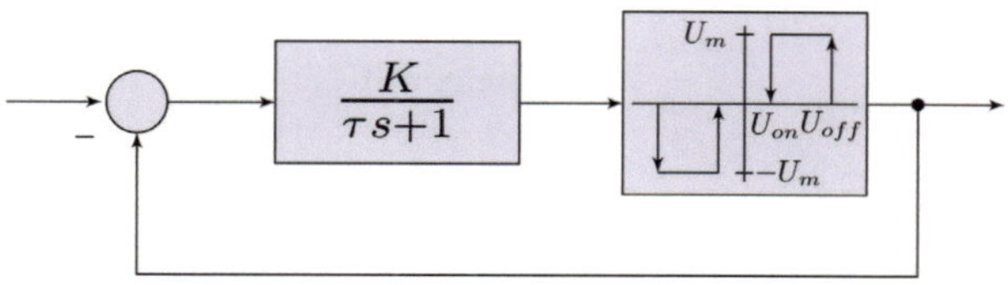

Figure 9.
Pulse-width pulse-frequency modulation.

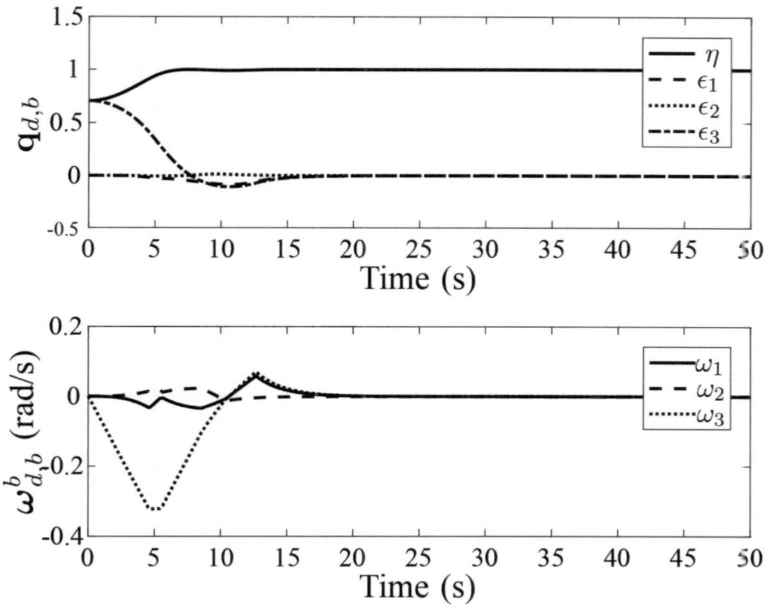

Figure 10.
Attitude control using thrusters with bang-bang modulation.

5. *Simulations of thruster modulations*: Consider a satellite with an inertia matrix as

$$\mathbf{J} = \begin{bmatrix} 0.5 & -0.2 & -0.1 \\ -0.2 & 0.5 & -0.2 \\ -0.1 & -0.2 & 0.5 \end{bmatrix}, \tag{31}$$

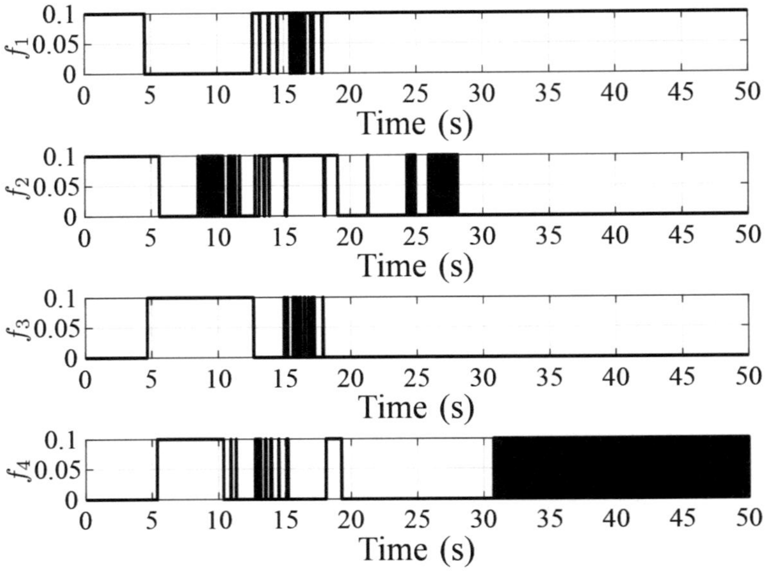

Figure 11.
Thruster firings when using bang-bang modulation.

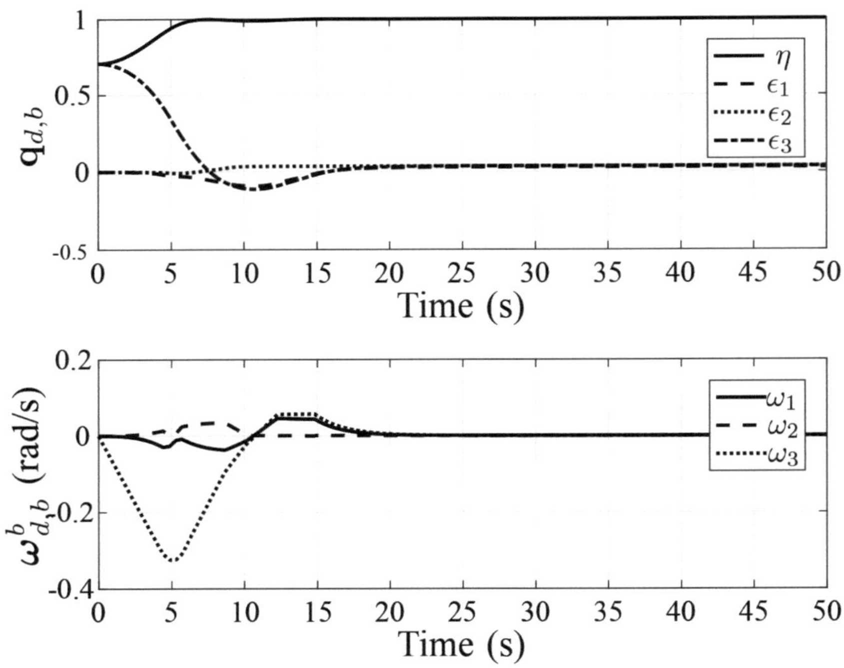

Figure 12.
Attitude control using thrusters with bang-bang modulation with dead zone.

where the objective is to perform a yaw maneuver of 90° using four thrusters with 0.1 N force, with a specific impulse of 200 s. **Figure 10** shows the attitude and angular velocity vectors when using bang-bang modulation, where the satellite is able to make the errors go to zero. However, due to the modulation, the thrusters will continue firing as shown in **Figure 11**. To that end, consider the bang-bang modulation with dead zone. Let the dead zone be chosen as $D = 0.05$, and then the satellite obtains an accuracy as shown in **Figure 12** where there is a small

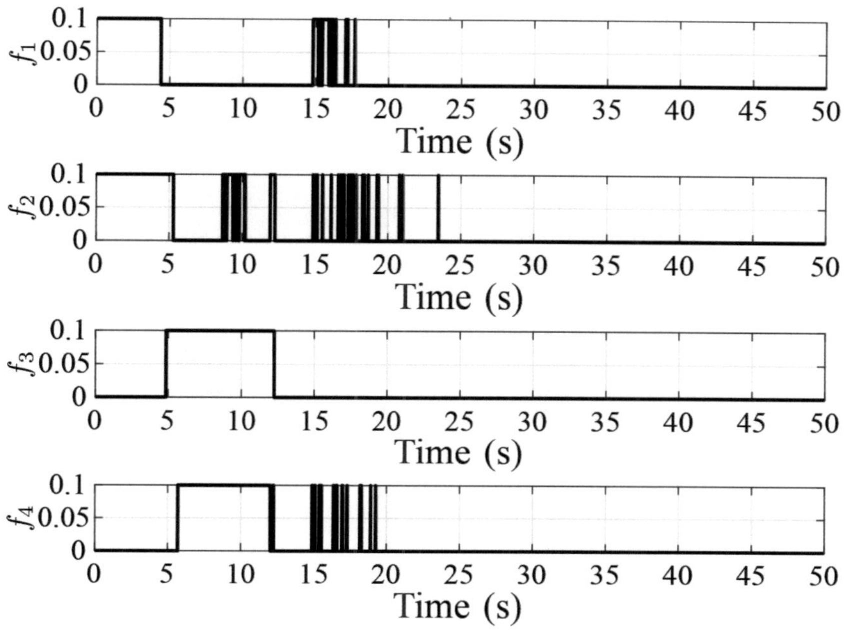

Figure 13.
Thruster firings when using bang-bang modulation with dead zone.

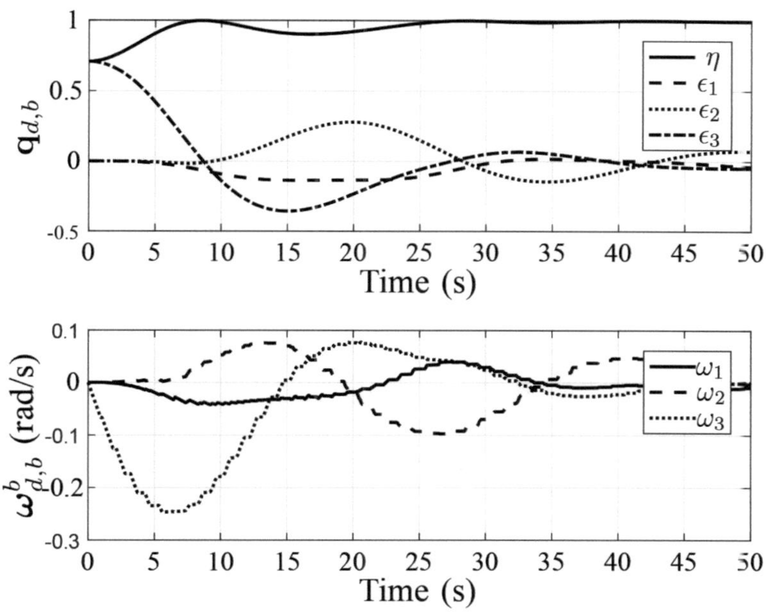

Figure 14.
Attitude control using thrusters with PWM modulation.

offset from the origin which is proportional to the dead zone. On the other hand, the thruster firings are much less prone to do continuous firings as shown in **Figure 13**.

Now, consider pulse-width modulation. Let the sawtooth signal have an amplitude of 1 and a frequency of 1 Hz. Then, the attitude and angular velocity error is obtained as shown in **Figure 14**, while the thruster firings are shown in **Figure 15**. It is possible to tune on sawtooth frequency to improve the performance.

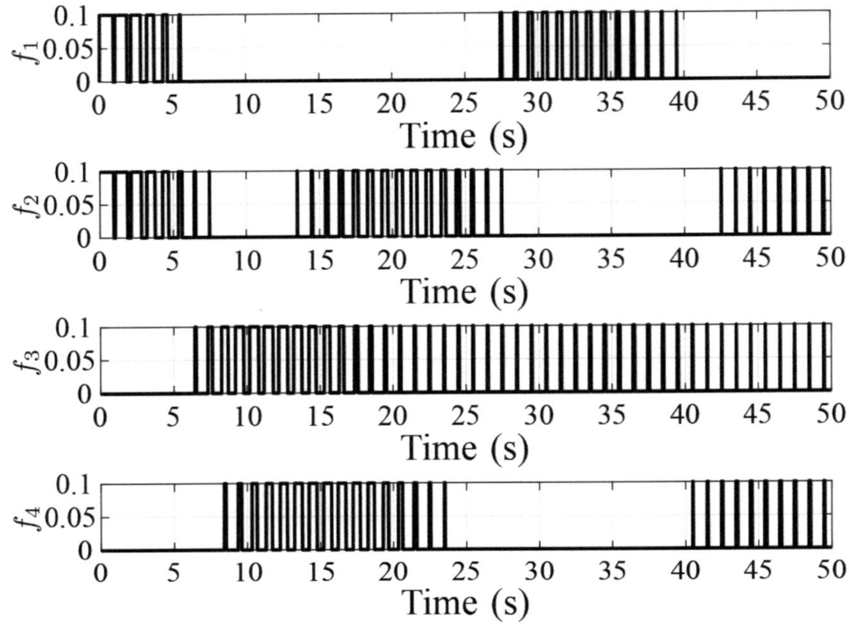

Figure 15.
Thruster firings when using PWM modulation.

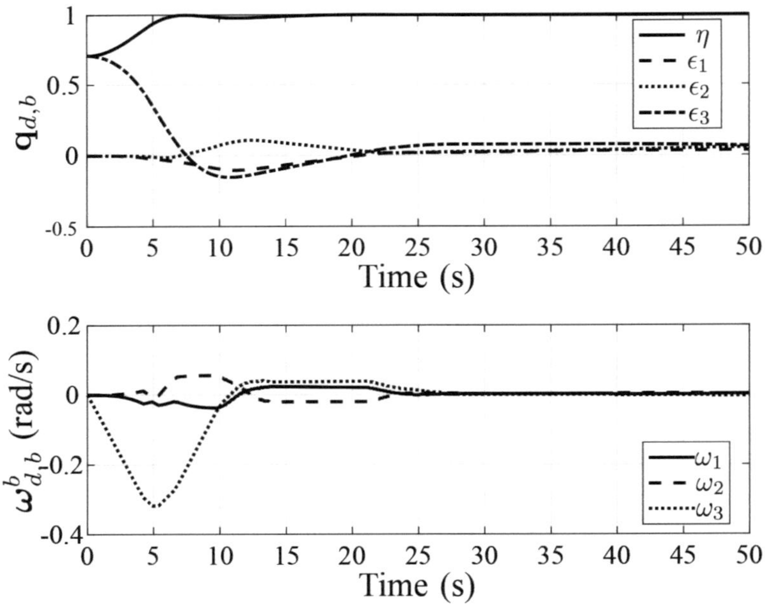

Figure 16.
Attitude control using thrusters with PWPF modulation.

The final scenario is using PWPF modulation, where the parameters are chosen as $K = 3$, $\tau = 0.2$, $U_{on} = 0.35$, and $U_{off} = 0.28$. **Figure 15** shows the attitude and angular velocity, which go close to zero, while the thruster firings are shown in **Figure 16**, which is able to constrain the amount of thruster firings, and therefore propellant.

For satellite control using thrusters, propellant is a critical resource that must not be wasted. To that end it is desirable to limit the amount of propellant while at the same time obtain good pointing accuracy (**Figure 17**). With the basis in

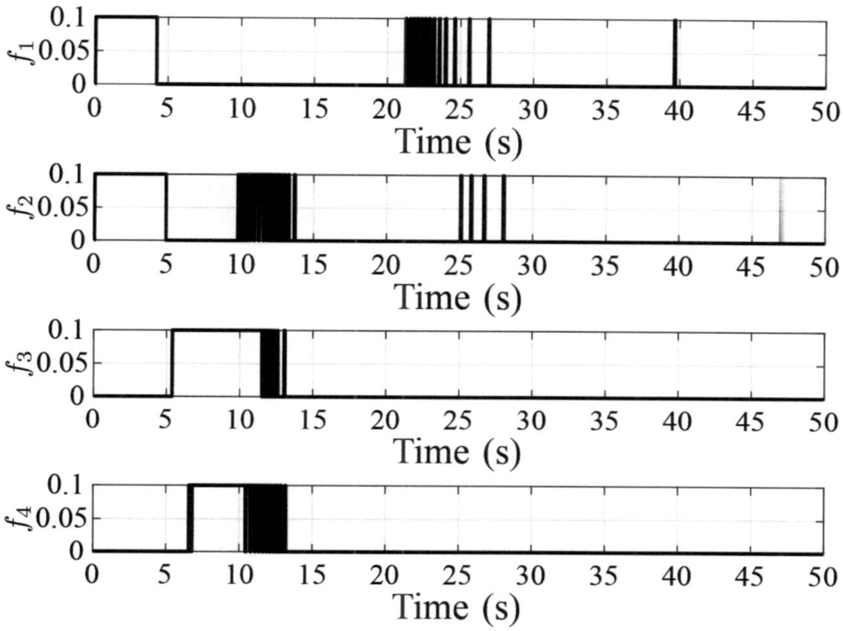

Figure 17.
Thruster firings when using PWPF modulation.

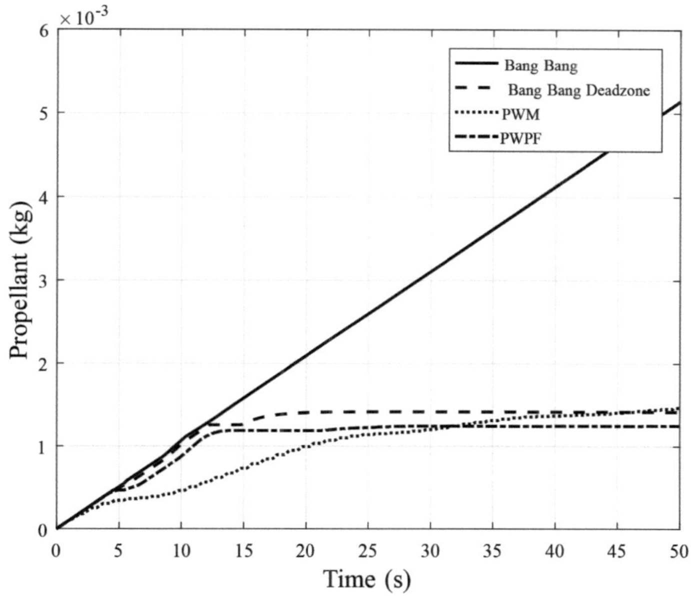

Figure 18.
Propellant consumption of the different modulation methods.

Tsiolkovsky's rocket equation, the propellant consumption during thruster firings can be found as $m_{propellant} = \int_0^t \frac{f}{I_{sp}g} dt$ where f is the force from one of the thrusters and I_{sp} is the specific impulse, while $g = 9.81$ m/s^2 is the acceleration due to gravity. **Figure 18** shows a comparison between the different modulation methods, where it is evident that the PWPF method allows for the least amount of propellant while obtaining close to acceptable performance. The bang-bang modulation will continue spending propellant until running out of fuel but on the other hand obtains the best tracking performance.

6. Attitude determination

As a preliminary step before trying to estimate the attitude of the satellite, some knowledge of measurement vectors must be known, i.e., what is the direction toward the Sun and how does the magnetic field vector look like at a given position. There are several other quantities that can be measured to obtain the attitude, where star trackers are known to be the most accurate. For the reader to obtain insight into using multiple measurements and combine it to find the attitude, this work presents a Sun vector model and a simplified magnetic field model that can be used for simulation purposes.

6.1 Sun vector model

To find the direction toward the Sun, there are several models that can be applied. The simplest would be to divide a circle into 365 days and have a vector always point toward the Sun. Then, by knowing which day it is, it is straightforward to find the direction toward the Sun. This approach would be coarse, such that more accurate models exist. For example, the Sun vector model in [3], pp. 281–282, has an accuracy of 0.01° and is valid until 2050. First, the time and date must be converted into the Julian date as [3], p. 189.

$$\begin{aligned} JD = 367(yr) - INT\left(\frac{7\left(yr + INT\left(\frac{mo+9}{12}\right)\right)}{4}\right) + INT\left(\frac{275mo}{9}\right) + d \\ + 1,721,013.5 + \frac{\frac{\left(\frac{s}{60^*}+\min\right)}{60} + h}{24}, \end{aligned} \tag{32}$$

where a real truncation is denoted by $INT()$ and the year, month, day, hour, minute, and second are denoted by $yr, mo, d, h, \min, s$. If the day contains a leap second, 61 s should be used instead of 60*. This gives the Sun vector model as

$$T_{UT1} = \frac{JD - 2,451,545.0}{36,525}, \tag{33}$$

$$\lambda_{M_\odot} = 280.460° + 36,000.771 T_{UT1}, \tag{34}$$

$$M_\odot = 357.5277233° + 35,999.05034 T_{UT1}, \tag{35}$$

$$\begin{aligned} \lambda_{ecliptic} = \lambda_{M_\odot} + 1.914666471° \sin(M_\odot) \\ + 0.019994643 \sin(2M) \end{aligned}, \tag{36}$$

$$\varepsilon = 23.439291° - 0.0130042 T_{UT1}, \tag{37}$$

$$\mathbf{s}^o = \mathbf{R}_i^o \begin{bmatrix} \cos\left(\lambda_{ecliptic}\right) \\ \cos\left(\varepsilon\right)\sin\left(\lambda_{ecliptic}\right) \\ \sin\left(\varepsilon\right)\sin\left(\lambda_{ecliptic}\right) \end{bmatrix}. \tag{38}$$

Here, the number of Julian centuries is denoted by T_{UT1}, the mean longitude of the Sun is denoted by $\lambda_{M_\odot}$, the mean anomaly for the Sun is denoted by $M_\odot$, the ecliptic longitude is denoted by $\lambda_{ecliptic}$, and the obliquity of the ecliptic is denoted by ε, while the Sun vector in orbit frame is denoted by $\mathbf{s}^o$.

6.2 Magnetic field model

Several different geomagnetic models can be applied for attitude determination in conjunction with a magnetometer. The most basic are simple dipole models [27], while more advanced are, e.g., the chaos model or the 12th generation IGRF model [28], which is the most commonly used model for attitude determination. This section presents the simple dipole model by [27], which can be described by the magnetic field vector in orbit frame as

$$\mathbf{m}^o = \frac{\mu_f}{a^3}\left[\,\cos\left(\omega_0 t\right)\sin\left(i\right) \quad -\cos\left(i\right) \quad 2\sin\left(\omega_0 t\right)\sin\left(i\right)\,\right]^{\mathrm{T}}, \tag{39}$$

where the time measured from passing the ascending node of the orbit relative to the geomagnetic equator is denoted by t and the dipole strength is denoted $\mu_f = 7.9 \cdot 10^{-15}$ Wb-m, while $\omega_0 = \|\omega_{i,o}^i\|$ denotes the angular speed of the orbit. For a real application, the reader is recommended to apply the IGRF model, which is available in Simulink inside the Aerospace Toolbox, as C++ implementation[2] or using Python.[3]

6.3 Attitude determination using the Madgwick filter

The objective of attitude determination is to find what direction the satellite is pointing. In its core, it mainly requires two vector measurements and two mathematical models that can be compared and used to find the attitude There are several different kinds of filters applied for attitude estimation, such as the Triad method [29], the Kalman filter [30], or the Mahony filter [31]. The Madgwick filter by [13] has shown good results in attitude estimation based on IMU measurements and is commonly applied in drone applications. The main idea by the filter is to use gradient descent in combination with the complementary filter to fuse sensor data together to produce an estimated quaternion. This section presents an application of the Madgwick filter by using measurements of the Sun vector and the magnetic field vector as well as the acceleration vector (gravity) and shows how to fuse that data together to estimate the attitude of a satellite in an elliptical orbit.

Let the quaternion estimate be denoted by $\hat{\mathbf{q}} = \begin{bmatrix} q_1 & q_2 & q_3 & q_4 \end{bmatrix}^{\mathrm{T}}$. The measured acceleration, Sun vector, and magnetic field vectors can be defined, respectively, as $\mathbf{a}^b$, $\mathbf{s}^b$, and $\mathbf{m}^b$ and can be combined with the mathematical models of the acceleration, Sun vector, and magnetic field vector given in Eqs. (6), (38), and (39) to estimate the attitude. Here, the current estimate is denoted by subscript k, while the previous estimate is denoted by $k-1$. Let the objective function be

[2] https://github.com/JDeeth/MagDec
[3] https://github.com/scivision/pyigrf12

$$\mathcal{F}=\begin{bmatrix}\mathbf{f}\left(\hat{\mathbf{q}}_{k-1},\mathbf{a}^{o},\mathbf{a}^{b}\right)\\ \mathbf{f}\left(\hat{\mathbf{q}}_{k-1},\mathbf{s}^{o},\mathbf{s}^{b}\right)\\ \mathbf{f}\left(\hat{\mathbf{q}}_{k-1},\mathbf{m}^{o},\mathbf{m}^{b}\right)\end{bmatrix}, \tag{40}$$

where the objective is found in an estimated quaternion that minimizes this function, something that can be achieved by using gradient descent. The Jacobian matrix can be found as

$$\mathcal{J}=\begin{bmatrix}\mathbf{J}_{q}\left(\hat{\mathbf{q}}_{k-1},\mathbf{a}^{o}\right)\\ \mathbf{J}_{q}\left(\hat{\mathbf{q}}_{k-1},\mathbf{s}^{o}\right)\\ \mathbf{J}_{q}\left(\hat{\mathbf{q}}_{k-1},\mathbf{m}^{o}\right)\end{bmatrix} \tag{41}$$

and allows the gradient to be found as $\nabla\,\mathbf{f}=\mathcal{J}^{\mathrm{T}}\,\mathcal{F}$. Now, let a vector in the orbit frame obtained from a mathematical model be denoted by $\mathbf{z}^{o}=\begin{bmatrix}o_x & o_y & o_z\end{bmatrix}^{\mathrm{T}}$ and a vector in the body frame obtained through measurement be denoted by $\mathbf{z}^{b}=\begin{bmatrix}b_x & b_y & b_z\end{bmatrix}^{\mathrm{T}}$. Then, the functions $\mathbf{f}\left(\hat{\mathbf{q}}_{k-1},\mathbf{z}^{o},,\mathbf{z}^{b}\right)$ and $\mathbf{J}_{q}\left(\hat{\mathbf{q}}_{k-1},\mathbf{z}^{o}\right)$ are given by

$$\mathbf{f}\left(\hat{\mathbf{q}}_{k-1},\mathbf{z}^{o},\mathbf{z}^{b}\right)=\begin{bmatrix}2o_x\left(0.5-q_3^2-q_4^2\right)+2o_y\left(q_1q_4+q_2q_3\right)+2o_z\left(q_2q_4-q_1q_3\right)-b_x\\ 2o_x\left(q_2q_3-q_1q_4\right)+2o_y\left(0.5-q_2^2-q_4^2\right)+2o_z\left(q_1q_2+q_3q_4\right)-b_y\\ 2o_x\left(q_1q_3+q_2q_4\right)+2o_y\left(q_3q_4-q_1q_2\right)+2o_z\left(0.5-q_2^2-q_3^2\right)-b_z\end{bmatrix}, \tag{42}$$

$$\mathbf{J}_{q}\left(\hat{\mathbf{q}}_{k-1},\mathbf{z}^{o}\right)=\begin{bmatrix}2o_yq_4-2o_zq_3 & 2o_yq_3+2o_zq_4 & -4o_xq_3+2o_yq_2-2o_zq_1 & -4o_xq_4+2o_yq_1+2o_zq_2\\ -2o_xq_4+2o_zq_2 & 2o_xq_3-4o_yq_2+2o_zq_1 & 2o_xq_2+2o_zq_4 & -2o_xq_1-4o_yq_4+2o_zq_3\\ 2o_xq_3-2o_yq_2 & 2o_xq_4-2o_yq_1-4o_zq_2 & 2o_xq_1+2o_yq_4-4o_zq_3 & 2o_xq_2+2o_yq_3\end{bmatrix}. \tag{43}$$

Given the gyro measurement $\boldsymbol{\omega}_{gyro}^{b}$ (relative to inertial frame), the angular velocity relative to orbit frame can be found as

$$\boldsymbol{\omega}_{o,gyro}^{b}=\begin{bmatrix}0\\ \boldsymbol{\omega}_{gyro}^{b}-\mathbf{R}_{o}^{b}\left(\hat{\mathbf{q}}_{k-1}\right)\mathbf{R}_{i}^{o}\boldsymbol{\omega}_{i,o}^{i}\end{bmatrix}\in\mathbb{R}^{4}, \tag{44}$$

where the rotation matrix from orbit to body frame is constructed using the estimated quaternion and denoted by $\mathbf{R}_{o}^{b}\left(\hat{\mathbf{q}}_{k-1}\right)$. The Madgwick filter can now be presented as

$$\hat{\boldsymbol{\omega}}_{k}^{b}=2\mathbf{T}\left(\hat{\mathbf{q}}_{k-1}^{*}\right)\frac{\nabla\mathbf{f}}{\|\nabla\mathbf{f}\|}, \tag{45}$$

$$\boldsymbol{\omega}_{bias,\,k}^{b}=\boldsymbol{\omega}_{bias,\,k-1}^{b}+\zeta\hat{\boldsymbol{\omega}}_{k}^{b}\Delta T, \tag{46}$$

$$\boldsymbol{\omega}_{o,\,b}^{b}=\mathbf{H}\left(\boldsymbol{\omega}_{o,gyro}^{b}-\boldsymbol{\omega}_{bias,\,k}^{b}\right), \tag{47}$$

$$\dot{\hat{\mathbf{q}}}_{k}=\frac{1}{2}\mathbf{T}\left(\hat{\mathbf{q}}_{k-1}\right)\begin{bmatrix}0\\ \boldsymbol{\omega}_{o,\,b}^{b}\end{bmatrix}-\beta\frac{\nabla\mathbf{f}}{\|\nabla\mathbf{f}\|}, \tag{48}$$

$$\hat{\mathbf{q}}_{k}=\hat{\mathbf{q}}_{k-1}+\dot{\hat{\mathbf{q}}}_{k}\Delta T, \tag{49}$$

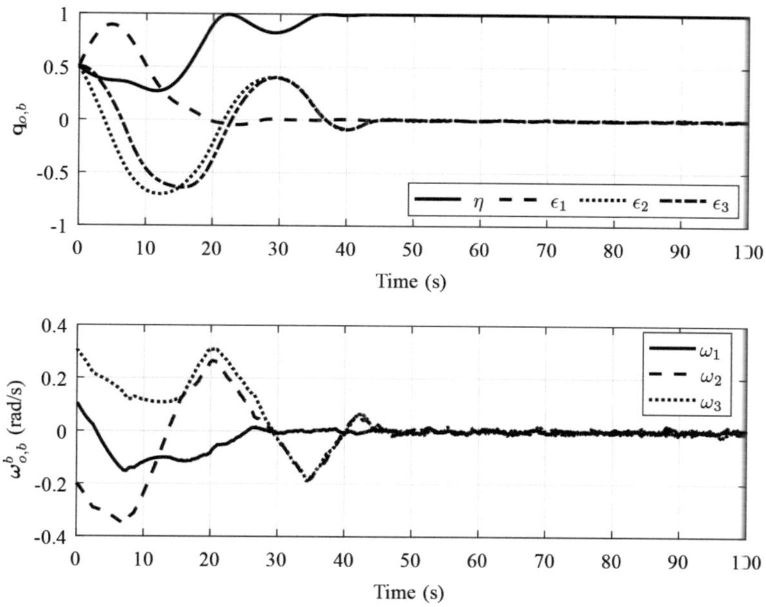

Figure 19.
Attitude and angular velocity during the maneuver [9].

$$\hat{\mathbf{q}}_k = \frac{\hat{\mathbf{q}}_k}{\|\hat{\mathbf{q}}_k\|}, \tag{50}$$

where β and ζ are gains, the time step is denoted by ΔT, the estimated angular velocity based on vector measurements is denoted $\hat{\omega}^b_k \in \mathbb{R}^4$, and the estimated gyro bias is denoted by $\boldsymbol{\omega}^b_{bias,k} \in \mathbb{R}^4$, while the angular velocity of the body frame relative to the orbit frame is denoted $\boldsymbol{\omega}^b_{o,b} \in \mathbb{R}^3$ (expected output) and the estimated quaternion is denoted $\hat{\mathbf{q}}_k$ describing the body frame relative to the orbit frame. Note that the quaternion must be normalized to ensure unit length and that the first elements of $\hat{\omega}^b_k, \boldsymbol{\omega}^b_{bias,k} \in \mathbb{R}^4$ are enforced to zero. To map the angular velocity from four to three elements, the projection matrix is defined as $\mathbf{H} = [\mathbf{0} \quad \mathbf{I}] \in \mathbb{R}^{3\times 4}$, which has a column vector of zeros followed by the identity matrix such that $\boldsymbol{\omega}^b_{o,b} \in \mathbb{R}^3$.

6.4 Simulation

Let a satellite have the inertia matrix as given in Eq. (43), which contains non-diagonal terms which therefore will create perturbing moments due to the gravity. Furthermore, let the satellite have the following initial conditions: $\mathbf{q}_{o,b}(0) = [0.5 \quad 0.5 \quad 0.5 \quad 0.5]^{\mathrm{T}}$ and $\boldsymbol{\omega}^b_{o,b} = [0.1 \quad -0.2 \quad 0.3]^{\mathrm{T}}$ with $\hat{\mathbf{q}}_{o,b}(0) = [1 \quad 0 \quad 0 \quad 0]^{\mathrm{T}}$ and $\hat{\omega}^b_{o,b} = [0 \quad 0 \quad 0]^{\mathrm{T}}$. The desired quaternion can be chosen as $\mathbf{q}_{o,d} = [1 \quad 0 \quad 0 \quad 0]^{\mathrm{T}}$, while $\boldsymbol{\omega}^d_{o,d} = \dot{\boldsymbol{\omega}}^d_{o,d} = \mathbf{0}$. To model noise in the sensor measurements, the quaternion is converted into Euler angles, where noise is added to the different sensors. Then, creating the rotation matrix from the noisy Euler angles allows the Sun vector, acceleration, and magnetometer models in the orbit frame to be rotated to the body frame, where the measurements now contain noise. The step size of the simulation is 0.01, while the sensors are sampled every 0.1 s.

The quaternion and angular velocity error of the satellite are shown in **Figure 18**. After about 50 s, the objective of making the attitude error and angular velocity error go to zero is achieved. Since the attitude is not measured directly, the Madgwick filter is used for attitude estimation as shown in **Figure 19**. Both the quaternion error (estimated truth) and angular velocity error converge close to zero.

From the PD+ controller, the desired torque is mapped to the desired thrust firings using bang-bang modulation (**Figure 20**). **Figure 21** shows the thruster firings required to maintain the attitude error close to zero.

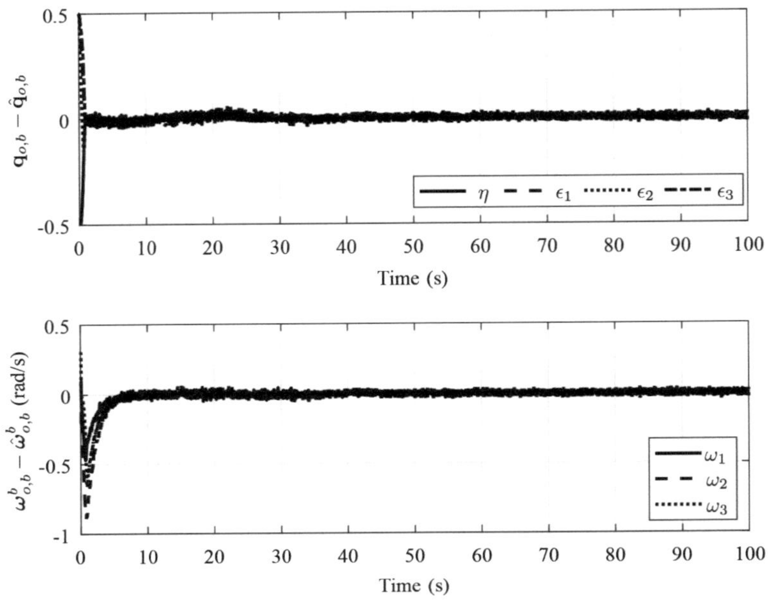

Figure 20.
Estimation error [9].

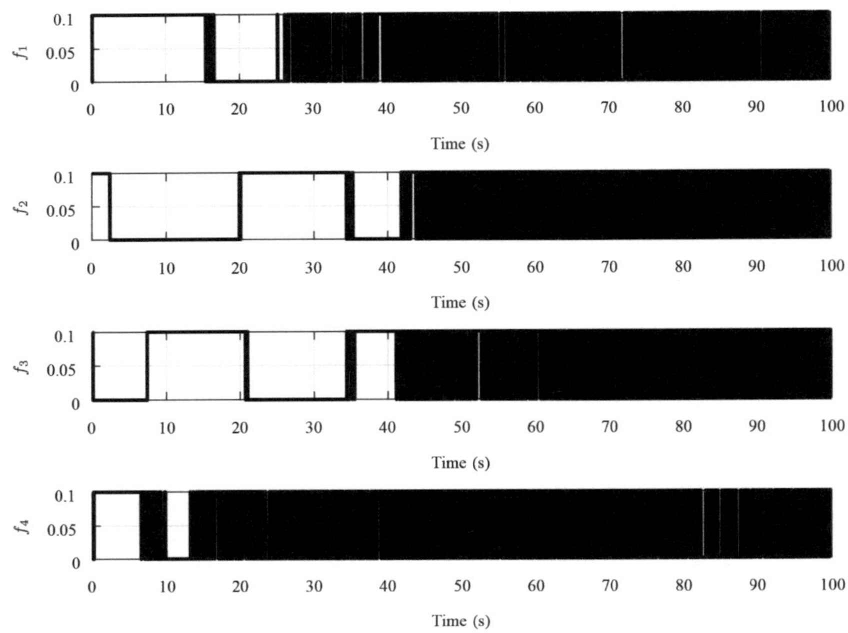

Figure 21.
Thruster firings to control the attitude [9].

7. Conclusion

This chapter has presented all the components required to create an attitude determination and control system for satellites in elliptical orbits. With this as basis, it is the hope by the author that the work can help in developing new results within attitude determination and control systems, ranging from nonlinear controllers to new understanding in orbital mechanics, attitude determination, new sensors, and new actuation methods and strategies.

Author details

Espen Oland
Department of Electrical Engineering, UiT - The Arctic University of Norway, Narvik, Norway

*Address all correspondence to: espen.oland@uit.no

References

[1] Sidi MJ. Spacecraft Dynamics & Control. New York: Cambridge University Press; 1997

[2] Wertz J, editor. Spacecraft Attitude Determination and Control. D. Dordrecht, Holland: Reidel Publishing Company; 1978

[3] Vallado DA. Fundamentals of Astrodynamics and Applications. 3rd ed. El Segundo, California: Microcosm Press and Dordrecht, Boston, London: Kluwer Academic Publishers; 2007

[4] Markley FL, Crassidis JL. Fundamentals of Spacecraft Attitude Determination and Control. El Segundo, California: Microcosm Press and New York, Heidelberg, Dordrecht London: Springer; 2014

[5] Alarcon J, Örger N, Kim S, Soon L, Cho M. Aoba VELOX-IV attitude and orbit control system design for a LEO mission applicable to a future lunar mission. In: Proceedings of the 67th International Astronautical Congress; Guadalajara City, Mexico; 2016

[6] Dechao R, Tao S, Lu C, Xiaoqian C, Yong Z. Attitude control system design and on-orbit performance analysis of nano-satellite - tian Tuo 1. Chinese Journal of Aeronautics. 2014;**27**(3): 593-601

[7] Nakajima Y, Murakami N, Ohtani T, Nakamura Y, Hirako K, Inoue K. SDS-4 attitude control system: In-flight results of three axis attitude control for small satellites. IFAC Proceedings Volumes. 2013;**16**(19):283-288

[8] Fritz M, Shoer J, Singh L, Henderson T, McGee J, Rose R, Ruf C. Attitude determination and control system design for the CYGNSS microsatellite. In: Proceedings of the IEEE Aerospace Conference; Big Sky, Montana; 2015

[9] Oland E. Attitude determination and control system for satellites in elliptical orbits—A complete solution. In: Proceedings of the 8th International Conference on Recent Advances in Space Technology (RAST); Istanbul, Turkey; 2017

[10] Oland E, Houge T, Vedal F. Norwegian student satellite program—HiNCube. In: Proceedings of the 22nd Annual AIAA/USU Conference on Small Satellites; Utah, USA; 2008

[11] Oland E, Schlanbusch R. Reaction wheel design for cubesats. In: Proceedings of the 4th International Conference on Recent Advances in Space Technologies (RAST); Istanbul, Turkey; 2009

[12] Schlanbusch R, Oland E, Nicklasson PJ. Modeling and simulation of a cubesat using nonlinear control in an elliptic orbit. In: Proceedings of the 4th International Conference on Recent Advances in Space Technologies (RAST); Istanbul, Turkey; 2009

[13] Madgwick S, Harrison A, Vaidyanathan R. Estimation of IMU and MARG orientation using a gradient descent algorithm. In: Proceedings of the 2011 IEEE International Conference on Rehabilitation Robotics; Zurich, Switzerland; 2011

[14] Oland E. Autonomous inspection of the International Space Station. In: Proceedings of the 8th International Conference on Mechanical and Aerospace Engineering; Prague, Czech; 2017

[15] Egeland O, Gravdahl JT. Modeling and Simulation for Automatic Control. Trondheim, Norway: Marine Cybernetics; 2002. ISBN: 82-92356-01-0

[16] Schaub H, Junkins JL. Analytical Mechanics of Space Systems. AIAA

American Institute of Aeronautics & Ast; 2003. pp. 1-19

[17] Oland E, Andersen TS, Kristiansen R. Subsumption architecture applied to flight control using composite rotations. Automatica. 2016;**69**:195-200

[18] Kristiansen R. Dynamic synchronization of spacecraft [PhD thesis]. Trondheim, Norway: Norwegian University of Science and Technology; 2008

[19] Takegaki M, Arimoto S. A new feedback method for dynamic control of manipulators. ASME Journal of Dynamic Systems, Measurement, and Control. 1981;**103**(2):119-125

[20] Paden B, Panja R. Globally asymptotically stable 'PD+' controller for robot manipulators. International Journal of Control. 1988;**47**(6): 1697-1712

[21] Kristiansen R, Nicklasson PJ, Gravdahl JT. Spacecraft coordination control in 6DOF: Integrator backstepping vs. passivity-based control. Automatica. 2008;**44**(11): 2896-2901

[22] Schlanbusch R, Loría A, Kristiansen R, Nicklasson PJ. PD+ based output feedback attitude control of rigid bodies. IEEE Transactions on Automatic Control. 2012;**57**(8):2146-2152

[23] Oland E. A command-filtered backstepping approach to autonomous inspections using a quadrotor. In: Proceedings of the 24th Mediterranean Conference on Control and Automation; Athens, Greece; 2016

[24] Hahn W. Stability of Motion. Berlin/Heidelberg/New York: Springer-Verlag; 1967. ISBN: 978-3-642-50087-9

[25] Wiśniewski R. Satellite attitude control using only electromagnetic actuation [PhD thesis]. Aalborg University; 1996

[26] Hu Q, Ma G. Flexible spacecraft vibration suppression using PWPF modulated input component command and sliding mode control. Asian Journal of Control. 2007;**9**(1):20-29

[27] Psiaki M. Magnetic torquer attitude control via asymptotic periodic linear quadratic regulation. Journal of Guidance, Control, and Dynamics. 2001;**24**(2):386-394

[28] Thébault E et al. International geomagnetic reference field: The 12th generation. Earth, Planets and Space. 2015;**67**(79):1-19

[29] Shuster MD, Oh SD. Three-axis attitude determination from vector observations. Journal of Guidance and Control. 1981;**4**(1):70-77

[30] Farrell JL. Attitude determination by kalman filtering. Automatica. 1970; **6**(3):419-430

[31] Mahony R, Hamel T, Pflimlin J. Nonlinear complementary filters on the special orthogonal group. IEEE Transactions on Automatic Control. 2008;**54**(5):1203-1217

Chapter 6

Data Mining-Based Identification of Nonlinear Systems

Natalia Bakhtadze, Vladimir Lototsky, Valery Pyatetsky and Alexey Lototsky

Abstract

This chapter presents identification methods using associative search of analogs and wavelet analysis. It investigates the properties of data mining-based identification algorithms which allow to predict: (i) the approach of process variables to critical values and (ii) process transition to chaotic dynamics. The methods proposed are based on the modeling of human operator decision-making. The effectiveness of the methods is illustrated with an example of product quality prediction in oil refining. The development of fuzzy analogs of associative identification models is further discussed. Fuzzy approach expands the application area of associative techniques. Finally, state prediction techniques for manufacturing resources are developed on the basis of binary models and a machine learning procedure, which is named associative rules search.

Keywords: process identification, knowledge base, associative search models, wavelet analysis

1. Introduction

The reduction of uncertainty in object description in terms of adjustable model has been a key conceptual direction in the identification theory and applications for a long time. In the statistical description of uncertainty, consistent estimates of plant's characteristics can be obtained by analyzing the convergence of the empirical distribution functional with the corresponding "theoretical" values, but this entails appropriate increase of the sample size. The difficulties in implementing this approach, especially for nonlinear and nonstationary objects, along with the increased possibilities of plant history analysis resulted in the advent of identification methods based on data mining [1].

The use of additional a priori information on the system for its training is considered by some authors today to be one of the key trends in the theory and practice of identification [2, 3].

One method that implements this approach to identification is the associative search method based on the design of predictive models [2]. They are based on inductive learning, that is, on associative search of analogs by means of intelligent analysis of process history and knowledge base development. The development of a predictive model for a dynamic object by associative search technique (i.e., by building a new model at every time step) is based on the generated and updated knowledge about the system. This approach allows to use any available a priori information about the plant [3].

The stability of a model built using the associative search techniques is investigated in terms of the spectrum analysis of a *multi-scale wavelet expansion* [4]. Methods based on the wavelet analysis open up a unique possibility to select "frequency-domain windows" as against the well-known windowed Fourier transform.

The development of intelligent identification algorithms for nonlinear and nonstationary objects is important for various applications, in particular, in chemical, oil refining, and power (smart grids) industries; transportation and logistics system; and trading processes (Bakhtadze et al. [1, 2, 4–7]).

2. Control system identification

Consider a traditional problem of dynamic object identification. For input vectors meeting Gauss-Markov assumptions, the least squares parameter estimates are consistent, unbiased, and efficient. However, the development of a closed-loop control system (for identification-based control system synthesis) faces considerable challenges. In a closed loop, the system state depends on control values at earlier time instants, which results in a degeneration problem.

To develop an informational model of control system's dynamics in a degenerate case, the Moore-Penrose method [8, 9] can be used for getting pseudo-solutions to a linear system by means of least squares techniques.

For a wide class of objects and, in particular, processes, control based on a linear model identification is not satisfactory. At the same time, models constructed by the method of associative search frequently are highly accurate even for nonlinear objects. However, some processes can be characterized by certain "irregularities" in certain time intervals, which affect the accuracy and adequacy of associative models.

Examples of such irregularities (which are often oscillatory in engineering systems) can be:

- seasonal and daily load oscillations in power networks that affect directly the optimization of power transmission control modes;
- ups and downs of stock market caused by various economic reasons;
- feed source changes in industrial process, and so on.

3. Associative search as intelligent modeling method

The difference between the associative search method based on data mining and traditional identification techniques is as follows. The method does not approximate process dynamics in time; it rather builds a new predictive model of the dynamic object (a "virtual model") at each time step using historical data sets ("associations") generated at the training phase.

As a result, at any time step, process control decision-making by a human individual (process operator, supervisor, plant or enterprise manager, trading operator, etc.) is modeled on the base of his/her knowledge and emerging associations.

Clustering (self-organizing learning) is an effective way to form associations.

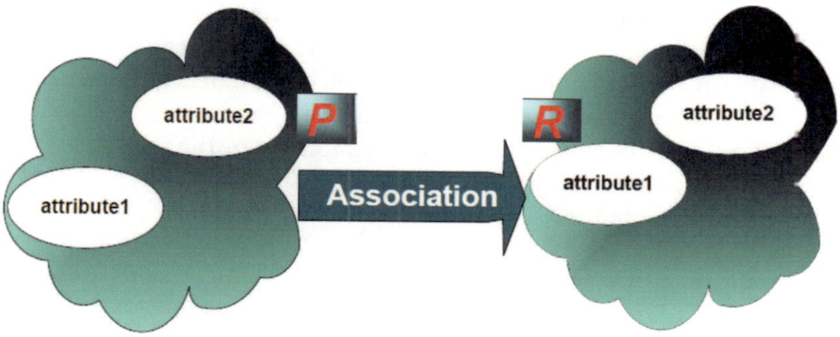

Figure 1.
Model of a human associative memory.

Knowledge in intelligent systems is of two types [10]. The first type of knowledge, that is, declarative knowledge, by means of appropriate ontologies describes different facts, events, and observation. A formal description of skills is called procedural knowledge. Depending on the level of this knowledge, users can be referred to as beginners or experts [11]. These two groups have different structures and ways of thinking. Beginners use so-called *inverse reasoning* in the procedure for decision-making. They make decisions based on the analysis of the information obtained in the previous step. In contrast to the beginners, experts at an intuitive, subconscious level form the so-called direct reasoning. Thus, cognitive psychology defines knowledge as a collection of symbols stored in the memory of a particular person [12]. The symbols, in turn, can be determined by their structure and the nature of neuron links [13].

Knowledge processing in an intelligent system consists in the recovery (associative search) of knowledge by its fragment [14]. The knowledge can be defined as an associative link between images (**Figure 1**). As an image, we will use "feature sets," that is, components of input vectors or input variables. The set of all associations over the set of images forms the memory of the intelligent system's knowledge base.

The associative search process can be either an image reconstruction procedure by a feature set (this set may not be complete; this approach is often used in models of a human associative memory), or the search procedure of other images in the archive, similar to the image under study by a certain criterion.

In Ref. [14], a model of decision-making search by the human operator is proposed, representing the process of associative thinking as a sequence of sets of associations. Association is a pair of images (the image-source and the image-output), wherein each image is described by a set of features. This approach is intermediate between neural networks and logical models in the classical theory of artificial intelligence.

The criterion for the similarity of two images in the general case can be represented as a logical function—a predicate. In the particular case, the features have a numerical expression. The feature sets that form the image are vectors in n-dimensional space. In this case, as a criterion of image similarity can be a metric in the space.

4. Associative search technique

Associative search method consists in constructing *virtual* predictive models. The term "virtual" should be understood as "ad hoc" [2]. The method presumes the construction of a predictive model for a dynamic object as follows. A traditional identification algorithm approximates real process in time. As against such algorithms, our method builds a new model at each time step t based on the analysis of the history data set ("associations") formed at the stage of learning and further adaptively corrected in accordance to certain criteria.

Within the present context, linear dynamic model is of the form:

$$y_N = \sum_{i=1}^{m} a_i y_{N-i} + \sum_{j=1}^{r_s} \sum_{s=1}^{S} b_{j,s} x_{N-j,s}, \forall j = 1,' N, \tag{1}$$

where y_N is the prediction of the object's output at the time instant N, x_N is the input vector, m is the memory depth in the output, r_s is the memory depth in the input, S is the dimension of the input vectors, and a_i and $b_{j,s}$ are tuning coefficients of the model. Model (1) is a regression whose structure is determined by a criterion of similarity of images forming the association.

In general, a new structure is formed for each time instant. The associative model is virtual in the sense that for each time step, it formed a new structure. For each current input vector, the corresponding input vectors and their corresponding outputs are selected from the archive. Further, a system of linear equations with respect to the adjustable coefficients is formed. Its decision in accordance with the least squares method determines the **point linear model of a nonlinear object,** as well as the output forecast.

Thus, each point of the global nonlinear regression surface is formed as a result of using linear "local" models at each new time step.

The set of values of inputs at each fixed point and the corresponding output replenish the procedural knowledge base.

Unlike classical regression models, for each fixed time instant from the process history, input vectors are selected close to the current input vector in the sense of a certain criterion (rather than the chronological sequence as in regression models). Thus, in Eq. (1), r_s is the number of vectors from the archive (from the time instant 1 to the time instant N), selected in accordance to the associative search criterion. A certain set of vectors r_s, $1 \le r_s \le N$, is selected at each time segment $[N-1, N]$. The criterion for selecting the input vectors from the archive is described below (**Figure 2**).

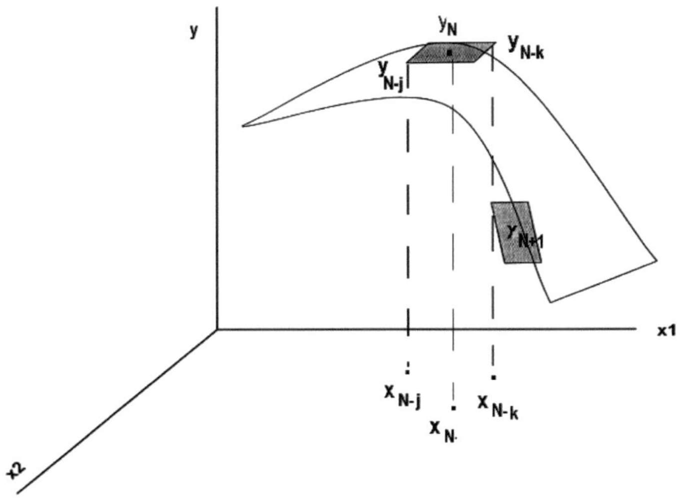

Figure 2.
Approximating hypersurface design.

As a distance (a norm in R^S) between points of the S-dimensional space of inputs, we introduce the value:

$$d_{N,N-j} = \sum_{s=1}^{S} \left| x_{N,s} - x_{N-j,s} \right|, \forall j = 1,'N, \tag{2}$$

where $x_{N,s}$ are the components of the input vector at the current time instant N. By virtue of a property of the norm ("the triangle inequality"), we have:

$$d_{N,N-j} \leq \sum_{s=1}^{S} \left| x_{N,s} \right| + \sum_{s=1}^{S} \left| x_{N-j,s} \right|, \forall j = 1,'N, \tag{3}$$

Let for the current input vector x_N:

$$\sum_{s=1}^{S} \left| x_{N,s} \right| = d_N. \tag{4}$$

To derive an approximating hypersurface for the vector x_N, we select from the archive of the input data such vectors $x_{N-j}, j = 1,'N$ that for a set D_N the condition:

$$d_{N,N-j} \leq d_N + \sum_{s=1}^{S} \left| x_{N-j,s} \right| \leq D_N, \forall j = 1,'N, \tag{5}$$

holds, where D_N may be selected, for instance, from the condition (**Figure 3**):

$$D_N \geq 2d_N^{max} = 2\max_j \sum_{s=1}^{S} \left| x_{N-j,s} \right|. \tag{6}$$

Under the assumptions that the inputs meet the Gauss-Markov conditions, the estimates obtained via the LS method are unbiased and statistically effective.

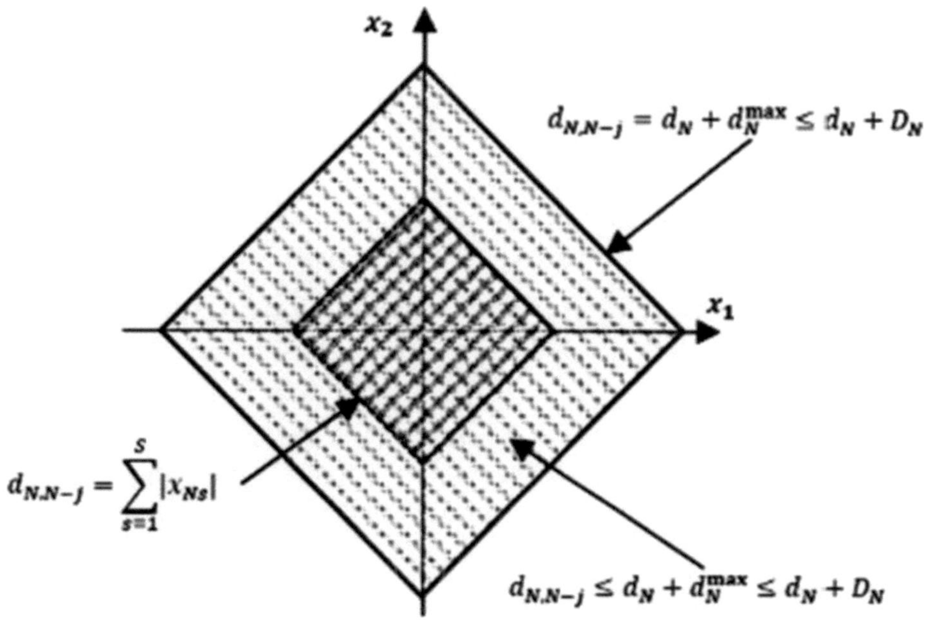

Figure 3.
Approximating hypersurface building.

5. Fuzzy virtual models

Fuzzy models under uncertainty are advisable to apply in decision-making systems in the following cases [3]:

- dynamics of the investigated quality index is described by a complex nonlinear dependence; and
- one or more factors of this dynamics are weakly or not formalized.

In fuzzy systems, the most commonly used technique is the production rule one. The production rule consists of antecedent (or several premises) and consequent. In the general case, the premises are connected by logical operators AND and OR.

Fuzzy systems are based on production-type rules with linguistic variables used as premise and conclusion in the rule.

By renaming the variables, the linear dynamic plant's model can be represented as follows:

$$Y_N = \sum_{i=1}^{n+\mathrm{T}} a_i X_i$$

The fuzzy system based on the production rules has the form:

A fuzzy model with $n+m$ input variables $\mathbf{X} = \{X_1, X_2, \ldots X_{n+m}\}$ defined in space $\mathrm{D}\mathbf{X} = \mathrm{D}X_1 \times \mathrm{D}X_2 \times \ldots \times \mathrm{D}X_{n+m}$ and with one-dimensional output Y is defined in the domain of reasoning DY.

Clear values of fuzzy variables X_i and Y are denoted by x_i and y, respectively.

$LX_i = \{LX_{i,1}, \ldots, LX_{i,l_i}\}$ is the fuzzy domain of definition of the i -th input variable and X_i is the number of linguistic terms on which this fuzzy variable is defined.

$LY = \{LY_1, \ldots, LY_{ly}\}$ is the domain of the fuzzy output variable.

l is the number of fuzzy values.

LY_j is the name of the output linguistic term.

The rule base in the fuzzy *Mamdani system* is a set of fuzzy rules such as:

$$R_j : \mathrm{LX}_{1,j_1} \text{ AND} \ldots \text{AND } \mathrm{LX}_{n,j_n} \rightarrow \mathrm{LY}_j. \tag{7}$$

The j-th fuzzy rule in the singleton-type system looks as follows:

$$R_j : \mathrm{LX}_{1,j_1} \text{ AND} \ldots \text{AND } \mathrm{LX}_{n,j_n} \rightarrow r_j \tag{8}$$

where r_j is a real number to estimate the output y.

The j-th rule in the Takagi-Sugeno model [15] looks as follows:

$$R_j : \mathrm{LX}_{1,j_1} \text{ AND} \ldots \text{AND } \mathrm{LX}_{n+m,j_{n+m}} \rightarrow r_{0j} + r_{1j}x_1 + r_{2j}x_2 + \ldots + r_{(n+m)j}x_{n+m} \tag{9}$$

where the output y is estimated by a linear function.

Thus, the fuzzy system performs the mapping $L : \mathfrak{R}^{n+m} \rightarrow \mathfrak{R}$.

The grade of crisp variable x_i membership in the fuzzy notion LX_{ij} is determined by membership functions $\mu_{LX_{ij}}(x_i)$. The rule base is formed by the criterion of minimum output error which can be defined by the following expressions:

$$\frac{\sum_{i=1}^{\mathrm{K}} |f(x_i) - L(x_i)|}{\mathrm{K}}, \frac{\sqrt{\sum_{i}^{\mathrm{K}} (f(x_i) - L(x_i))^2}}{\mathrm{K}}, \max_{i \in \mathrm{K}} |f(x_i) - L(x_i)| \tag{10}$$

where K is the number of samples.

Depending on the features of the object and the purpose of identification, various fuzzy models can be formed. Thus, the Takagi-Sugeno model is most suitable for objects with complex nonlinear dynamics, such as moving objects, in the control of which the accuracy requirements prevail.

A fuzzy model of the Mamdani type is suitable for problems in the solution of which it is important to form knowledge based on data analysis.

The singleton-type system may be used in both identification and knowledge-formation tasks.

Singleton-type fuzzy model performs the mapping $L : \mathfrak{R}^{n+m} \to \mathfrak{R}$ where the fuzzy conjunction operator is replaced by a product, and the operator of fuzzy rules aggregation, that is, by summation. The mapping L is defined by the following expression:

$$L(x) = \frac{\sum_{i=1}^{q} \mu_{\mathrm{LX}_{1i}}(x_1) \cdot \mu_{\mathrm{LX}_{2i}}(x_2) \cdot \ \dots \ \cdot \mu_{\mathrm{LX}_{(n+m)i}}(x_{n+m}) \cdot r_i}{\sum_{i=1}^{q} \mu_{\mathrm{LX}_{1i}}(x_1) \cdot \mu_{\mathrm{LX}_{2i}}(x_2) \cdot \ \dots \ \cdot \mu_{\mathrm{LX}_{(n+m)i}}(x_{n+m})} \tag{11}$$

where $\mathbf{x} = [x_1, \dots, x_{n+m}]^T \in \mathfrak{R}^{n+m}$; q is the number of rules in a fuzzy model; $n + m$ is the number of input variables in the model; and $\mu_{\mathrm{LX_{ij}}}(x_{ij})$ is the membership function.

The expression for L mapping in the Takagi-Sugeno model looks as follows:

$$L(x) = \frac{\sum_{i=1}^{q} \mu_{\mathrm{LX}_{1i}}(x_1) \cdot \mu_{\mathrm{LX}_{2i}}(x_2) \cdot \ \dots \ \cdot \mu_{\mathrm{LX}_{(n+m)i}}(x_{n+m}) \cdot \left(r_{0i} + r_{1i}x_1 + r_{2i}x_2 + \ \dots \ + r_{(n+m)i}x_{n+m}\right)}{\sum_{i=1}^{q} \mu_{\mathrm{LX}_{1i}}(x_1) \cdot \mu_{\mathrm{LX}_{2i}}(x_2) \cdot \ \dots \ \cdot \mu_{\mathrm{LX}_{(n+m)i}}(x_{n+m})} \tag{12}$$

In Mamdani fuzzy systems, fuzzy logic techniques are used for describing the input vector's $\boldsymbol{x}$ mapping into the output value y, for example, Mamdani approximation or a method based on a formal logical proof.

Let the variables in (1) be fuzzy. In this case, (1) can be represented as a fuzzy model of Takagi-Sugeno (TS) [15].

To form the model, product rules with linear finite-difference equations on the right-hand side are defined (for simplicity, we consider one-input case, i.e., $P = 1$):

If $y(t-1)$ is Y_1^θ,..., $y(t-r)$ is Y_r^θ,
$x(t)$ is X_0^θ,..., $x(t-s)$ is X_r^θ, then

$$y^\theta(t) = a_0^\theta + \sum_{k=1}^{r} a_k^\theta y(t-k) + \sum_{l=0}^{s} b_l^\theta x(t-l), \theta = 1, \dots, n, \tag{13}$$

where: $a^\theta = \left(a_0^\theta, a_1^\theta, \dots, a_r^\theta\right)$, $b^\theta = \left(b_0^\theta, b_1^\theta, \dots, b_s^\theta\right)$ are adjustable parameter vectors; $\boldsymbol{y}(t-r) = (1, y(t-1), \dots, y(t-r))$ is the state vector; $\boldsymbol{x}(t-s) = (x(t), x(t-1), \dots, x(t-s))$ is an input sequence; and $Y_1^\theta, \dots, Y_r^\theta, X_0^\theta, \dots, X_r^\theta$ are fuzzy sets.

By re-denoting input variables: $(u_0(t), u_1(t), \dots, u_m(t)) = (1, y(t-1), \dots, y(t-r), x(t), \dots x(t-s))$, finite-difference equation's coefficients: $\left(c_0^\theta, c_1^\theta, \dots, c_m^\theta\right) = \left(a_0^\theta, a_1^\theta, \dots, a_r^\theta, b_1^\theta, \dots, b_s^\theta\right)$, and membership functions:

$$\left(U_1^\theta(u_1(t)), \dots, U_m^\theta(u_m(t))\right) = \left(Y_1^\theta(y(t-1)), \dots, Y_r^\theta(y(t-r)), X_0^\theta(x(t)), \dots, X_s^\theta(x(t-s))\right),$$

where $m = r = s + 1$, one obtains the analytic form of the fuzzy model, intended for calculating the output $\hat{y}(t)$:

$$\hat{y}(t) = c^T \widetilde{u}(t), \tag{14}$$

where $c = \left(c_0^1, \ldots, c_0^n, \ldots, c_m^1, \ldots, c_m^n\right)^T$ is the vector of the adjustable parameters;

$\widetilde{u}T(t) = \left(u_0(t)\beta^1(t), \ldots, u_0(t)\beta^\theta(t), \ldots, u_m(t)\beta^1(t), \ldots, u_m(t)\beta^n(t)\right)$ is the extended input vector;

$$\beta^\theta(t) = \frac{U_1^\theta(u_1(t)) \otimes \ldots \otimes U_m^\theta(u_m(t))}{\sum_{\theta=1}^{N}\left(U_1^\theta(u_1(t)) \otimes \ldots \otimes U_m^\theta(u_m(t))\right)} \tag{15}$$

is a fuzzy function where $\otimes$ denotes the minimization operation of fuzzy product.

If for $t = 0$, the vectorc$(0) = 0$, the correcting $mn \times nm$ matrix $Q(0)$ (m is the number of input vectors, n is the number of production rules), and the values of $u(t), t = 1, \ldots, N$ are specified, the parameter vector $c(t)$ is calculated using the known multi-step LSM:

$$\begin{aligned} c(t) &= c(t-1) + Q(t)\widetilde{u}(t)\left[y(t) - c^T(t-1)\widetilde{u}(t)\right] \\ Q(t) &= Q(t-1) - \frac{Q(t-1)\widetilde{u}(t)\widetilde{u}^T(t)Q(t-1)}{1 + \widetilde{u}T(t)Q(t-1)\widetilde{u}(t)} \end{aligned} \tag{16}$$

$Q(0) = \gamma \mathrm{I}, \gamma > > 1,$ where I is the unit matrix.

The above equations show that even in case of one-dimensional input and few production rules, a lot of observations are needed to apply LSM which makes the fuzzy model too unwieldy. Therefore, only a part of the whole set of rules ($r < n$) should be chosen according to a certain criterion.

The application of the associative search techniques where one or more model parameters are fuzzy is reduced to such determination of the predicate $\Xi = \left\{\Xi_i\left(R_0^a, \mathrm{R}^a, \mathrm{T}^a\right)\right\}$, so that the number of production rules in the TS model is significantly reduced according to some criterion.

For example, the following matrix:

$$\begin{matrix} \beta_1^{\Theta_t} & \ldots & \beta_P^{\Theta_t} \\ \ldots & \ldots & \ldots \\ \beta_1^{\Theta_{t-s}} & \ldots & \beta_P^{\Theta_{t-s}} \end{matrix} \tag{17}$$

can be defined for P-dimensional input vectors at time steps $t{-}j, j = 1, \ldots, s$. If the rows of this matrix are ranged, say, w.r.t. $\sum_{p=1}^{P}\left|\beta_p^{\Theta_i}\right|$ decrease and a certain number of rows are selected, then such selection combined with condition (4) will determine the predicate Ξ and, respectively, the criterion for selecting the images (sets of input vector) from the history.

Let us range the rows of this matrix, for example, subject to the criterion of descending the values $\sum_{p=1}^{P}\left|\beta_p^{\Theta_i}\right|$, and select a certain number of rows. Such selection combined with condition (4) defines the predicate $\Xi = \left\{\Xi_i\left(R_0^a, \mathrm{R}^a, \mathrm{T}^a\right)\right\}$, and, respectively, the image selection criterion (sets of input vectors) from the archive.

5.1 Fuzzy associative search

Notwithstanding all benefits delivered by fuzzy techniques, their application significantly reduces the calculation speed that is critical for predicting the dynamics of some plants. This consideration coupled with the principal impossibility of formalizing some factors necessitated the development of algorithms that could combine all advantages of fuzzy approach and associative search algorithms.

Assume the associative search procedure is determined by the predicate $\Xi(P^a, R^a)$, which interprets input variables' limits (specified, say, by process specifications) as a fuzzy conjunction of input variables:

$$\Xi(P^a, R^a) = \{(X_1 : x_1 \subset A_1) \wedge (X_2 : x_2 \subset A) \dots (X_n : x_n \subset A_n\}$$

for all $X_1, X_2, \dots X_n$ from $DX = DX_1 \times DX_2 \times \dots \times DX_n$.

Then, the production rules, where fuzzy variables possess such values that $\Xi(P^a, R^a)$ possesses the value FALSE, will be discarded automatically. This reduces drastically the number of production rules employed in the fuzzy model and thus increases significantly the algorithms' speed.

6. Solving the associative search problem by means of clusterization techniques

The associative search problem is solved by clustering technique (both crisp and fuzzy) in the following way.

The current vector under investigation is attributed to a certain cluster per the criterion of minimum distance to the center:

$$\min_k \sum_{k=1}^{K} \left\| g_k - \acute{x}_N \right\|^2,$$

where $\acute{x}_N \in X$ is the current input vector of the control plant under investigation.

Within this cluster, the vectors are sought that satisfy the assigned associative criterion. It may turn out that one cannot find within this cluster the number of vectors necessary to solve the problem of forecasting using the method of least squares. In this case, one of the known methods of combining two clusters with the minimum distance between any two of their members can be applied. This approach provides significant savings in computing resources compared to searching through a full search. However, such a combination of clusters does not yet guarantee the solution of the problem. The approach described below looks the most reasonable.

6.1 Virtual clustering ("impostor" method)

The current input vector at any particular time can be assigned to a specific cluster. This can, for example, be done by the criterion of the minimum distance to the center.

Let

$$\min_k \sum_{k=1}^{K} \left\| g_k - \acute{x}_N \right\|^2$$

be satisfied for $k = r$.

Let $\acute{x}_N$ denote the center of the cluster $\mathbf{A}_r$. If additional selection of input vectors from the archive is required (to form a system of a sufficient number of equations to identify the system using the associative search method), clusters with the minimum distance between their centers and $\acute{x}_N$ are selected for the join. This approach allows not only to discard a significant number of vectors removed from $\acute{x}_N$, but also to select from the archive the maximum possible number of vectors satisfying the criterion of associative search.

After the completion of this procedure, assigning $\acute{x}_N$ as the cluster center $\mathbf{A}_r$ is canceled, and the procedure of the formation of virtual (relevant to the certain time instant) models continues using conventional clustering algorithms.

7. Case study: oil refining product quality modeling

Key process equipment of an atmospheric distillation unit comprises of cold and hot crude oil preheat trains, desalter, a flash drum or, instead, a pre-flash column with an overhead reflux drum, atmospheric heaters, and an atmospheric distillation column with a reflux drum and three side stripping columns for middle distillates (typically, kerosene, light diesel and heavy diesel aka atmospheric gas oil). The naphtha streams from both reflux drums are re-combined and further sent to downstream stabilization and rerun facilities. The atmospheric residuum from the bottom of the main atmospheric column is typically streamed to a vacuum distillation section.

To obtain a soft sensor model for the 10% distillation point of a kerosene stream, the lab data for this quality were collected along with process data from the atmospheric column. The predictive model is formed by means of the associative search method. The process data were analyzed, and process variables measured by plant instruments were selected for modeling along with the distillation point sampled at the plant and measured in the refinery's laboratory. Based on the preliminary data analysis, the following linear predictive model was developed:

$$T(t) = \sum_{i=1}^{4} b_i F_i(t-1) + b_5 F_5(t-3) + b_6 F_6(t-5) + \sum_{i=7}^{12} b_i F_i(t-7), \tag{18}$$

where $T(t)$ is the desired estimate; $F_i(t-j)$ are various process parameters, such as flows, temperatures, and pressures, measured directly at the plant; and $b_1, \ldots, b_{12}$ are model's coefficients.

The forecast was calculated per linear and associative models for 10,525 time steps (1 step = 10 min). **Figure 4** shows simulation results for the steps $t = 102^{'}, 301$.

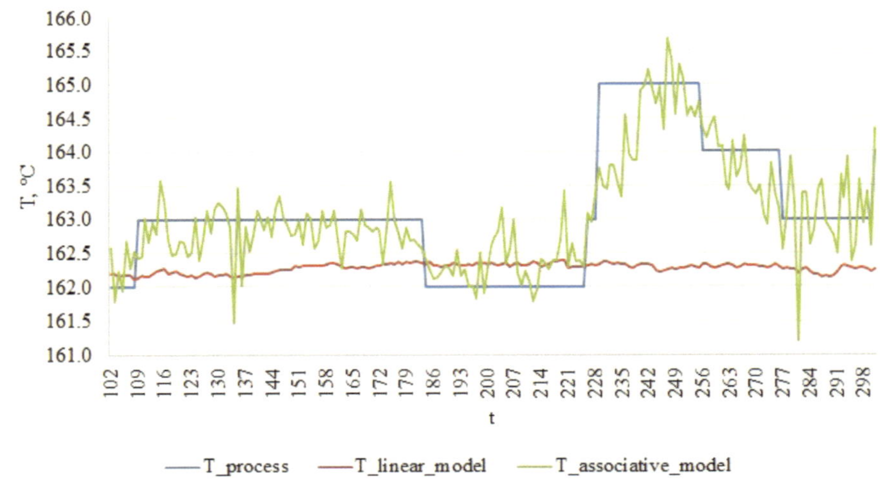

Figure 4.
Kerosene 10% distillation point forecast.

8. Application of wavelet approach to the analysis of nonstationary processes

Within the last two decades, applying wavelet transform (WT) to the analysis of nonstationary processes has been widely used. The wavelet transform of signals is a generalization of the spectral analysis, for instance, with regard to the Fourier transform.

First papers on the wavelet analysis of time (spatial) series with a pronounced heterogeneity appeared in the end of 1980s [16, 17]. The method was positioned as an alternative to the Fourier transform, localizing the frequencies but not providing the time extension of a process under study. In sequel, the theory of wavelets has appeared and is developed, as well as its numerous applications.

The scope of wavelet analysis today is very wide: it includes the synthesis and processing of nonstationary signals, compression and coding of information, image recognition and image analysis, the study of functions and time-dependent signals and inhomogeneity in space. The approach is effective for tasks where the results of the analysis should contain not only the characteristics of the frequency signal (signal power distribution by frequency components) but also information about local coordinates in which certain groups of frequency components manifest themselves or in which rapid changes in the frequency components of the signal occur. A significant number of practical applications have been created, including in health care, the study of geophysical fields, temporary meteorological series, and prediction of earthquakes [18].

The wavelet analysis method consists in applying a special linear conversion of signals. In particular, it becomes possible to study the physical properties or dynamics of real objects and processes in depth. For example, it can be processes in manufacturing. The wavelet transform (WT) of a one-dimensional signal is its representation in the form of a generalized Fourier series (or Fourier integral) over a system of basis functions called the "wavelet." A wavelet is characterized by the fact that the function that forms it (a wavelet-formation function or a wavelet matrix) is distinguished by a certain scale (frequency) and localization in time based on the time shift and the change in the time scale.

The time scale is analogous to the oscillation period, that is, it is inverse one with regard to the frequency, and the shift interprets the displacement of the signal over the time axis.

The wavelet transform performs the projection of a one-dimensional process into a two-dimensional surface in three-dimensional space. The frequency and time are treated as independent variables.

At the same time, it becomes realistic to simultaneously study the properties of the process being studied both in the time domain and in the frequency domain. It becomes possible to investigate the dynamics of the frequency process and its local features. This allows us to identify the coordinates at which certain frequencies manifest themselves most significantly.

The graphical representation of the wavelet analysis can be displayed in the form of isolines, illustrating the change in the intensities of wavelet transform coefficients at different time scales, and also for revealing local extrema of surfaces.

If a function is used in the Fourier transform that generates an orthonormal basis of space by means of a scale transformation, then the wavelet transform is formed using a basis function localized in a bounded domain, although defined on the whole numerical axis.

The wavelet transform, as a mathematical tool, serves mainly to analyze data in the time and frequency domains.

Wavelet transformation, as a mathematical tool, provides the ability to analyze data in the time and frequency domains simultaneously. The wavelet transform can provide time-frequency information about a function that in many practical situations is more relevant than information obtained through standard Fourier analysis.

There are examples of the use of wavelet analysis in identification problems [5]. In the literature, it is noted that wavelets are used mainly to identify nonlinear systems with a certain structure, where unknown time-varying coefficients can be represented as a linear combination of basis wavelet functions [6, 7]. It was stated that along with the usual ("direct") wavelet analysis, biorthogonal bursts [18], wavelet frames [19], or wavelet networks [20] can be used to identify the system.

There exist many different ways of applying wavelets for linear system identification. In Ref. [21], the identification of systems with a specific input/output structure was studied, in which the parameters are identified via spline-wavelets and their derivatives. In paper [22], an extended use of an orthonormal transformation least squares method is presented in order to reveal useful information from data.

9. Conditions of the associative model stability in the aspect of the analysis of the spectrum of multi-scale wavelet expansion

Let (1) be an associative search model. We represent the multi-scale wavelet decomposition for the current input vector $x(t)$ for a fixed level of detail L [7]:

$$\begin{aligned} x(t) &= \sum_{k=1}^{N} c^{x}_{L,k}(t)\varphi_{L,k}(t) + \sum_{l=1}^{L}\sum_{k=1}^{N} d^{x}_{l,k}(t)\psi_{l,k}(t), \\ y(t) &= \sum_{k=1}^{N} c^{y}_{L,k}(t)\varphi_{L,k}(t) + \sum_{l=1}^{L}\sum_{k=1}^{N} d^{y}_{l,k}(t)\psi_{l,k}7(t), \end{aligned} \tag{19}$$

where L is the depth of the multi-scale expansion; $\varphi_{L,k}(t)$ are scaling functions; $\psi_{l,k}(t)$ are the wavelet functions that are obtained from the mother wavelets by tension/combustion and shift

$$\psi_{l,k}(t) = 2^{l/2}\psi_{\text{mother}}\left(2^{l}t - k\right)$$

(as the mother wavelets, in the present case, we consider the Haar wavelets); l is the level of data detailing; $c_{L,k}$ are the scaling coefficients; and $d_{l,k}$ are the detailing coefficients. The coefficients are calculated by use of the Mallat algorithm [17].

Let us expand Eq. (1) over wavelets:

$$\begin{aligned} &\sum_{k=1}^{N} c^{y}_{Lk}(t)\varphi_{Lk}(t) + \sum_{l=1}^{L}\sum_{k=1}^{N} d^{y}_{lk}(t)\psi_{lk}(t) = \sum_{k=1}^{N}\left(\sum_{i=1}^{m} a_i c^{y}_{Lk}(t-i)\varphi_{Lk}(t-i)\right) \\ &+ \sum_{l=1}^{L}\sum_{k=1}^{N}\left(\sum_{i=1}^{m} a_i d^{y}_{lk}(t-i)\psi_{lk}(t-i)\right) + \sum_{k=1}^{N}\left(\sum_{s=1}^{S}\sum_{j=1}^{r_s} b_{sj} c^{s}_{Lk}(t-j)\varphi_{Lk}(t-j)\right) \\ &+ \sum_{l=1}^{L}\sum_{k=1}^{N}\left(\sum_{s=1}^{S}\sum_{j=1}^{r_s} b_{sj} d^{s}_{lk}(t-j)\psi_{lk}(t-j)\right) \end{aligned}$$

Let us consider individually the detailing and approximating parts correspondingly:

$$(t)\psi_{lk}(t) = \sum_{i=1}^{m} a_i d^{y}_{lk}(t-i)\psi_{lk}(t-i) + \sum_{s=1}^{S}\sum_{j=1}^{r_s} b_{sj} d^{s}_{lk}(t-j)\psi_{lk}(t-j), \tag{20}$$

$$c^{y}_{Lk}(t)\varphi_{Lk}(t) = \sum_{i=1}^{m} \hat{a}_i c^{y}_{Lk}(t-i)\varphi_{Lk}(t-i) + \sum_{s=1}^{S}\sum_{j=1}^{r_s} \hat{b}_{sj} c^{s}_{Lk}(t-j)\varphi_{Lk}(t-j). \tag{21}$$

In [7], it was shown that a sufficient condition for the stability of plant (1) is as follows: for $\forall k = \overline{1, N}$ meeting the inequalities is to be provided:

1. if $m > R$, $R = \max_{s=1,S} r_s$, then the condition for the detailing coefficients:

$$\left|\frac{a_1 d^y_{l,k}(t-1) + \sum_{s=1}^{S} b_{s,1} d^{x_s}_{l,k}(t-1)}{2d^y_{l,k}(t)}\right| < 1,$$
$$\left|\frac{-a_2 d^y_{l,k}(t-2) + \sum_{s=1}^{S} b_{s,2} d^{x_s}_{l,k}(t-2)}{a_1 d^y_{l,k}(t-1) + \sum_{s=1}^{S} b_{s,1} d^{x_s}_{l,k}(t-1)}\right| < 1, \dots,$$
$$\left|\frac{-a_{R+1} d^y_{l,k}(t-R-1)}{a_R d^y_{l,k}(t-R) + \sum_{s=1}^{S} b_{s,R} d^{x_s}_{l,k}(t-R)}\right| < 1, \tag{22}$$
$$\left|\frac{-a_{R+2} d^y_{l,k}(t-R-2)}{a_{R+1} d^y_{l,k}(t-R-1)}\right| < 1, \dots,$$
$$\left|\frac{-2a_m d^y_{l,k}(t-m)}{a_{m-1} d^y_{l,k}(t-m+1)}\right| < 1$$

for the approximating coefficients:

$$\left|\frac{a_1 c^y_{L,k}(t-1) + \sum_{s=1}^{S} b_{s,1} c^{x_s}_{L,k}(t-1)}{2c^y_{L,k}(t)}\right| < 1,$$
$$\left|\frac{-a_2 c^y_{L,k}(t-2) + \sum_{s=1}^{S} b_{s,2} c^{x_s}_{L,k}(t-2)}{a_1 c^y_{L,k}(t-1) + \sum_{s=1}^{S} b_{s,1} c^{x_s}_{L,k}(t-1)}\right| < 1, \dots,$$
$$\left|\frac{-a_{R+1} c^y_{L,k}(t-R-1)}{a_R c^y_{L,k}(t-R) + \sum_{s=1}^{S} b_{s,R} c^{x_s}_{L,k}(t-R)}\right| < 1, \tag{23}$$
$$\left|\frac{-a_{R+2} c^y_{L,k}(t-R-2)}{a_{R+1} c^y_{L,k}(t-R-1)}\right| < 1, \dots,$$
$$\left|\frac{-2a_m c^y_{L,k}(t-m)}{a_{m-1} c^y_{L,k}(t-m+1)}\right| < 1;$$

2. if $m < R$, $R = \max r_{s\, s=\overline{1,S}}$, then the condition for the detailing coefficients:

$$\left|\frac{a_1 d^y_{l,k}(t-1) + \sum_{s=1}^{S} b_{s,1} d^{x_s}_{l,k}(t-1)}{2d^y_{\mathrm{lk}}(t)}\right| < 1,$$
$$\left|\frac{-a_2 d^y_{l,k}(t-2) + \sum_{s=1}^{S} b_{s,2} d^{x_s}_{l,k}(t-2)}{a_1 d^y_{l,k}(t-1) + \sum_{s=1}^{S} b_{s,1} d^{x_s}_{l,k}(t-1)}\right| < 1, \dots,$$
$$\left|\frac{-\sum_{s=1}^{S} b_{s,m+1} d^{x_s}_{l,k}(t-m-1)}{a_m d^y_{l,k}(t-m) + \sum_{s=1}^{S} b_{s,m} d^{x_s}_{l,k}(t-m)}\right| < 1,$$
$$\left|\frac{-\sum_{s=1}^{S} b_{s,m+2} d^{x_s}_{l,k}(t-m-2)}{\sum_{s=1}^{S} b_{s,m+1} d^{x_s}_{l,k}(t-m-1)}\right| < 1, \dots,$$
$$\left|\frac{-2\sum_{s=1}^{S} b_{s,R} d^{x_s}_{l,k}(t-R)}{\sum_{s=1}^{S} b_{s,R-1} d^{x_s}_{l,k}(t-R+1)}\right| < 1$$

for the approximating coefficients:

$$\left|\frac{a_1c^y_{L,k}(t-1)+\sum_{s=1}^{S}b_{s,1}c^{x_s}_{L,k}(t-1)}{2c^y_{L,k}(t)}\right|<1,$$
$$\left|\frac{-a_2c^y_{L,k}(t-2)+\sum_{s=1}^{S}b_{s,2}c^{x_s}_{L,k}(t-2)}{a_1c^y_{L,k}(t-1)+\sum_{s=1}^{S}b_{s,1}c^{x_s}_{L,k}(t-1)}\right|<1,\ \dots,$$
$$\left|\frac{-\sum_{s=1}^{S}b_{s,m+1}c^{x_s}_{L,k}(t-m-1)}{a_mc^y_{L,k}(t-m)+\sum_{s=1}^{S}b_{s,m}c^{x_s}_{L,k}(t-m)}\right|<1, \tag{24}$$
$$\left|\frac{-\sum_{s=1}^{S}b_{s,m+2}c^{x_s}_{L,k}(t-m-2)}{\sum_{s=1}^{S}b_{s,m+1}c^{x_s}_{L,k}(t-m-1)}\right|<1,\ \dots,$$
$$\left|\frac{-2\sum_{s=1}^{S}b_{s,R}c^{x_s}_{L,k}(t-R)}{\sum_{s=1}^{S}b_{s,R-1}c^{x_s}_{L,k}(t-R+1)}\right|<1;$$

3. if $m=R\neq 1, R=maxr_{s\,s=1,'S}$, then the condition of the stability for the detailing coefficients:

$$\left|\frac{a_1d^y_{l,k}(t-1)+\sum_{s=1}^{S}b_{s,1}d^{x_s}_{l,k}(t-1)}{2d^y_{l,k}(t)}\right|<1,$$
$$\left|\frac{-a_2d^y_{l,k}(t-2)+\sum_{s=1}^{S}b_{s,1}d^{x_s}_{l,k}(t-2)}{a_1d^y_{l,k}(t-1)+\sum_{s=1}^{S}b_{s,1}d^{x_s}_{l,k}(t-1)}\right|<1,\ \dots, \tag{25}$$
$$\left|\frac{-2\left[a_md^y_{l,k}(t-m)+\sum_{s=1}^{S}b_{s,m}d^{x_s}_{l,k}(t-m)\right]}{a_{m-1}d^y_{l,k}(t-m+1)+\sum_{s=1}^{S}b_{s,m-1}d^{x_s}_{l,k}(t-m+1)}\right|<1$$

for the approximating coefficients:

$$\left|\frac{a_1c^y_{L,k}(t-1)+\sum_{s=1}^{S}b_{s,1}c^{x_s}_{L,k}(t-1)}{2c^y_{L,k}(t)}\right|<1,$$
$$\left|\frac{-a_2c^y_{L,k}(t-2)+\sum_{s=1}^{S}b_{s,2}c^{x_s}_{L,k}(t-2)}{a_1c^y_{L,k}(t-1)+\sum_{s=1}^{S}b_{s,1}c^{x_s}_{L,k}(t-1)}\right|<1,\ \dots, \tag{26}$$
$$\left|\frac{-2\left[a_mc^y_{L,k}(t-m)+\sum_{s=1}^{S}\hat{b}_{s,m}c^{x_s}_{L,k}(t-m)\right]}{a_{m-1}c^y_{L,k}(t-m+1)+\sum_{s=1}^{S}\hat{b}_{s,m-1}c^{x_s}_{L,k}(t-m+1)}\right|<1;$$

4. if $m=R=1, R=maxr_{s\,s=1,'S}$, then the condition of the stability for the detailing coefficients:

$$\left|\frac{a_1d^y_{l,k}(t-1)+\sum_{s=1}^{S}b_{s,1}d^{x_s}_{l,k}(t-1)}{d^y_{l,k}(t)}\right|<1$$

for the approximating coefficients:

$$\left| \frac{a_1 c^{y}_{L,k}(t-1) + \sum_{s=1}^{S} b_{s,1} c^{x_s}_{L,k}(t-1)}{c^{y}_{L,k}(t)} \right| < 1.$$

10. Prediction of the transfer to chaos

The chaotic system dynamics is characterized by considerable dependence on initial conditions, when as close as needed at the initial time instant trajectories during certain time are diverge by a finite distance. The main characteristics of the chaotic behavior are the speed of divergence of the trajectories defined by the senior Lyapunov exponent. This speed is determined by the Lyapunov exponent whose value represents the degree of instability or degree of sensitivity to the original data.

For a linear system with a constant matrix, the senior Lyapunov exponent is $\chi_1 = max\ \Re\lambda_i$, where λ_i are the eigenvalues of the system matrix. In other words, $|\chi_1|$ coincides with the conventional degree of the system stability [23].

Thus, (23) and (24) are sufficient conditions for chaotic dynamics prediction, what is a key condition under implementing phase transfers of technological processes under study.

11. Prediction of manufacturing situations

Optimal routine enterprise resource planning and scheduling are currently based on detailed mathematical models of production processes [24]. Rescheduling requires model update subject to the current production information.

Present-day industrial sites feature interrelated multi-variable production processes and sophisticated material flow networks; scheduling at such sites poses nonlinear NP-hard optimization problems.

The state of manufacturing resources should be nevertheless assessed and predicted both to improve control agility and to foresee the situations where schedule execution becomes problematic or impossible. Such situations will be further referred to as incidents.

It may make sense to develop intelligent predictive models describing the overall current state of resources employed to execute all production operations of a specific production process.

The term "production resources" will hereafter mean the following:

- input flows characterized by formal properties dependent on production specificity; and

- production equipment.

$$d_{ij}, i = 1, \ldots, N; j = 1, \ldots, M$$

and other facilities used for performing the j-th operation;

- human resources

$$h_{ij}, i = 1, \ldots, H; j = 1, \ldots, M$$

involved in the j-th operation;

- other factors

$$f_{ij}, k = 1, \ldots, N; j = 1, \ldots, M$$

affecting the j-th operation such as energy resources and a variety of formal indices and factors related with the production process.

Production resources may be described differently.

1. Some have qualitative characteristics which take on specific values that may be checked against norms at any moment.

2. The state of others such as certain equipment pieces may be exclusively either "working" or "not working." The remaining life time may be known or not for such resources. The process historian may however keep failure statistics for a specific equipment piece; maintenance downtime statistics may be also available for a specific piece or similar kind of equipment.

3. One more resource type (including human resources) is not subject to maintenance. In case of outage, such resources should be immediately replaced from the backlog. The replacement process is typically fast; therefore, no values other than 1 (OK) and 0 (not OK) should be assigned to such resource.

Assume a model of a specific manufacturing situation as a dynamic schedule fragment comprising the following components:

$$r_{ij}(t) = \{ < C_1 > C_2 > C_3 > C_4 > C_5 > \}_{ijt} \tag{27}$$

where

$< C_1 > \stackrel{\text{def}}{=} < ijt >$ is a *resource identifier* including the resource number, the operation number, and the time stamp (the number of characteristics may be increased).

Other components of the resource state vector at the time moment t may be represented by a binary code.

$< C_2 >$ is the code of the numerical value of a state variable; this code is different for each of the above-listed resource types.

$< C_3 >$, $< C_4 >$, and $< C_5 >$ will be discussed further.

Consider the resources whose state may be described by some quantitative characteristic, such as inlet flow rate or temperature for chemical processes or an average equipment failure number.

For a specific resource, we assume that the characteristic of its state possesses the values on the half-interval $[0; 1)$ (this half-interval was chosen as an example for simplicity, the results can be easily spread to any other).

This half-interval can be represented as the union

$[0; 0.5) \cup [0.5, 1)$. We will further correspond the symbols $\{0; 1\}$ to the left and right half-intervals respectively, namely, 0 to the left half-interval, and 1 to the right one.

Each of the two subintervals can be further split in the same way, and, again, the values 0 and 1 can be assigned to the left and the right parts, respectively.

In that way, a finite chain of symbols from $\{0; 1\}$ has a one-to-one correspondence with a half-interval embedded in $[0; 1)$. For a binary partition, a chain of n symbols corresponds to a half-interval with the length $\frac{1}{2^n}$.

This way, for each value of a numerical characteristic at the current time moment, we obtain a code of 0s and 1s. The number of positions, as we show further, will determine the accuracy of prediction.

For the resources from the categories 2 and 3, the respective codes will have the same value in all positions (either 1 or 0). $< C_3 >$ is the code of the time before the maintenance end. If a resource is available and operated, the respective code consists of 1s. $< C_4 >$ is the code of the time before the equipment piece fails with the probability close to 1 (remaining life).

In the scheduling practice, this time is not less than the operating time. However, resource replacement just during the operation may be sometimes more cost-effective. Moreover, the equipment piece may fail unexpectedly. For resource types from categories 1 and 3, $< C_4 >$ has 1s in all positions.

$< C_5 >$ is the time before the scheduled end of the operation. In real-life manufacturing situations, time may be wasted (with the need in schedule update) for the reasons neither stipulated in the production model nor caused by equipment failures.

Generally, it is hardly possible to formalize all such causes of schedule disruption. Therefore, their consolidation as the "remaining plan execution time" is a way to allow for these hidden factors in the production state model.

For the developed binary chain, a forecast may be obtained using data mining techniques. It makes sense to apply the methods named **association rules search** [25].

A forecast of a state described by a binary chain with an identifier can be obtained by revealing the most probable combination of two binary sets of values at a fixed time instant and at the next instant (a one-step forecast). A more distant prediction horizon is also possible.

12. Conclusion

Modern information technologies offer new possibilities for solving identification problems for control and decision-making systems. Data mining methods allow to solve problems that in the general case could not be solved by classical methods, or required heuristic approaches.

In this chapter, associative search techniques are presented. The techniques allow the identification of nonlinear systems, without the need to build a bunch of Wiener-Hammerstein models, etc. An alternative is to analyze the current state of the system using the knowledge base and training system. This approach allows the best use of a priori information on the object.

The algorithms may be successfully applied in the identification of nonlinear nonstationary processes. For these purposes, the multi-scale wavelet expansion is used. By investigating the dynamics of the coefficients of this expansion, one can predict the approach of process parameters to stability limits. Finally, sufficient conditions of stability are derived.

The high accuracy of forecasting by associative search technique makes it relevant for studying the dynamics of processes and predicting the transition to chaos. Also, it becomes possible to predict the contingencies of production processes. For this, the method of searching for associative rules is applied.

Author details

Natalia Bakhtadze[1,2]*, Vladimir Lototsky[1], Valery Pyatetsky[3] and Alexey Lototsky[1]

1 V.A. Trapeznikov Institute of Control Sciences, Moscow, Russia

2 Bauman Moscow State Technical University, Moscow, Russia

3 National University of Science and Technology "MISIS", Moscow, Russia

*Address all correspondence to: sung7@yandex.ru

References

[1] Peretzki D, Isaksson A, Carvalho A, Bittencourt C, Forsman K. Data Mining of Historic Data for Process Identification. Sweden: Linköping University Electronic Press; 2014. http://manualzz.com/doc/8482583/modeling-and-diagnosis-of-friction-and-wear-in-industrial

[2] Bakhtadze N, Kulba V, Lototsky V, Maximov E. Identification-based approach to soft sensors design. IFAC-PapersOnLine. 2007;**10**:302-307. DOI: 10.3182/20100701-2-PT-4011.00052

[3] Bakhtadze N, Maximov E, Valiakhmetov R. Fuzzy soft sensors for chemical and oil refining processes. IFAC Proceedings Volumes. 2008;**41**: 4246-4250. DOI: 10.3182/20080706-5-KR-1001.00017

[4] Bakhtadze N, Lototsky V, Vlasov S, Sakrutina E. Associative search and wavelet analysis techniques in system identification. IFAC Proceedings Volumes. 2012;**45**:1227-1232. DOI: 10.3182/20120711-3-BE-2027.00242

[5] Bakhtadze N, Sakrutina A. The intelligent identification technique with associative search. International Journal of Mathematical Models and Methods in Applied Sciences. 2015;**9**:418-431. ISSN: 1998-0140

[6] Bakhtadze N, Lototsky V. Knowledge-based models of nonlinear systems based on inductive learning. In: New Frontiers in Information and Production Systems Modelling and Analysis Incentive Mechanisms, Competence Management, Knowledge-based Production. Heidelberg: Springer; 2016. pp. 85-104. ISBN 978-3-319-23338-3. DOI: 10.1007/978-3-319-23338-3

[7] Bakhtadze N, Sakrutina E, Pyatetsky V. Predicting oil product properties with intelligent soft sensors. IFAC-PapersOnLine. 2017;**50**:14632-14637. DOI: 10.1016/j.ifacol.2017.08.1742

[8] Moore E. On the reciprocal of the general algebraic matrix. Bulletin of the American Mathematical Society. 1920; **26**:394-395

[9] Penrose R. A generalized inverse for matrices. Proceedings of the Cambridge Philosophical Society. 1955;**51**(3): 406-413

[10] Larichev OI, Asanov A, Naryzhny Y, Strahov S. Expert system for the diagnostics of acute drug poisonings, applications and innovations in intelligent systems IX. In: Macintosh A, Moulton M, Preece A, editors. Proceedings of the 21 SGES International Conference on Knowledge Based Systems and Applied Artificial Intelligence. Cambridge, UK: Springer-Verlag; 2001. pp. 159-168

[11] Patel V, Ramoni M. Cognitive models of directional inference in expert medical reasoning. In: Feltovich P, Ford K, Hofman R, editors. Expertise in Context: Human and Machine. Menlo Parc, CA: AAAI Press; 1997

[12] Hunt E. Cognitive science: Definition, status and questions. Annual Review of Psychology. 1989;**40**:603-629

[13] Newell A, Simon HA. Human Problem Solving. Englewood Cliffs, NJ: Prentice-Hall Inc.; 1972

[14] Gavrilov A. The model of associative memory of intelligent system. In: Proceedings of 6-th Russian-Korean International Symposium on Science and Technology. Novosibirsk. Vol. 1. 2002. pp. 174-177

[15] Takagi T, Sugeno M. Fuzzy identification of systems and its

applications to modeling and control. IEEE Transactions on Systems, Man, and Cybernetics. 1985;**26**:116-132

[16] Daubechies I, Lagarias J. Two-scale difference equations I: Existence and global regularity of solutions. SIAM Journal on Mathematical Analysis. 1991; **22**:1388-1410

[17] Mallat S. In: Barlaud M, editor. Wavelet Tour of Signal Processing. San Diego; CA: Academic Press; 1999. 635p

[18] Váňa Z, Preisig H. System identification in frequency domain using wavelets: Conceptual remarks. Systems & Control Letters. 2012;**61**(10): 1041-1051

[19] Ho K, Blunt S. Adaptive sparse system identification using wavelets. IEEE Trans. Circuits and Systems-II: Analog and Digital Signal Processing. 2003;**49**(10):656-667

[20] Sureshbabu N, Farrell JA. Wavelet-based system identification for nonlinear control. IEEE Transactions on Automatic Control. 1999;**44**(2):412-417

[21] Preisig HA. Parameter estimation using multi-wavelets. Computer Aided Chemical Engineering. 2010;**28**:367-372

[22] Carrier J, Stephanopoulos G. Wavelet-based modulation in control-relevant process identification. AICHE Journal. 1998;**44**(2):341-360

[23] Fradkov A, Evans R. Control of chaos: Survey—1997–2000. IFAC Proceedings Volumes. 2002;**35**:131-142

[24] Al-Otabi GA, Stewart MD. Simulation model determines optimal tank farm design. Oil & Gas Journal. 2004;**102**(7):50-55

[25] Qin JS, Badgwell TA. A survey of industrial model predictive control technology. Control Engineering Practice. 2003;**11**(7):733-764

Chapter 7

Analysis and Control of Power Electronic Converters Based on a System Zero Locations Approach

Jorge-Humberto Urrea-Quintero, Nicolás Muñoz-Galeano and Lina-María Gómez-Echavarría

Abstract

This chapter presents a procedure to design and control power electronic converters (PECs), which includes a zero-based analysis as a dynamical system response criterion for dimensioning converter passive elements. For this purpose, a nonideal boost DC-DC converter (converter considering its parasitic losses) is dynamically modeled and analyzed in steady state as an application example. The steady-state model is obtained from the average nonlinear model. The steady-state model allows deducing expressions for equilibrium conversion ratio $M(D)$ and efficiency η of the system. Conditions for the converter conduction modes are analyzed. Simulations are made to see how parasitic losses affect both $M(D)$ and η. Then, inductor current and capacitor voltage ripple analyses are carried out to find lower boundaries for inductor and capacitor values. The values of the boost DC-DC converter passive elements are selected taking into account both steady-state and zero-based analyses. A nonideal boost DC-DC converter and a PI-based current mode control (CMC) structure are designed to validate the proposed procedure. Finally, the boost DC-DC converter is implemented in PSIM and system operating requirements are satisfactorily verified.

Keywords: power electronic converters, boost DC-DC converter, zero-based analysis, current mode control, parasitic loss analysis, efficiency

1. Introduction

Design procedures of PECs must establish a trade-off between passive elements' values and dynamical performance because of the close dependence between them. Dynamical performance should not be deteriorated and operating requirements must be satisfied [1]. This task generally implies the construction of a nonlinear dynamical model and its implementation in any computational tool [2].

Dynamical modeling and steady-state analyses of PECs have received significant attention as tools to model system design [3]. Through dynamical modeling, it is possible to perform an analysis of the system behavior and its relation with passive elements' values [1]. Meanwhile, steady-state analysis provides expressions to determine in PEC: (a) $M(D)$, (b) η, and (c) continuous conduction mode (CCM) and discontinuous conduction mode (DCM) boundaries [3].

Multi-resolution PEC models can be constructed where parasitic losses can be taken into account [4]. However, if parasitic losses are not considered, the PEC model is simplified; but models do not adequately represent the PEC behavior in its entire operation range [5]. Moreover, a simplified model cannot predict both $M(D)$ and η nonlinearities and limitations [6].

Parasitic losses are typically modeled as appropriate equivalent series resistances (ESRs) associated with passive elements of PECs [3, 7, 8]. Parasitic losses can be included in the PEC design stage when both dynamical performance and η are taken into account [6]. Several works, [4, 7] to mention some of them, propose different PEC modeling approaches that have included parasitic losses. Nevertheless, in the reviewed literature, a consensus about what is the suitable detailed level of the model does not exist, in which PEC's dynamical behavior can be accurately represented; without the model, deduction becomes a challenge for the designer. However, the trend remains with the so-called average models which describe low-frequency and neglect high-frequency dynamics (semiconductor switching dynamics) of the system [9].

Average models that take, some or all, parasitic losses into account, have been presented by [1, 10]. Recent works [7, 10–14] show that a practical level of model detail for PECs includes parasitic losses associated with their passive elements and disregards losses due to semiconductor switching. Models with this level of detail are suitable for system design, $M(D)$ derivation, η analysis, and dynamical performance evaluation [12]. Additionally, these models are suitable for control purposes [2, 10].

It is clear that based on average models, PECs can be designed to carry out dynamical performance analysis. Notwithstanding, a design procedure is needed that comprises all necessary steps to design and control PECs and fulfills all given operating requirements. This design procedure must be simple and useful.

In the PEC field, few works that take into account dynamical characteristics of the system have been carried out [15–17]. In these works, PEC's design problem is presented as an optimization problem. In consequence, a procedure to easily design and control PECs is still needed. In this chapter, a procedure to easily design and control PECs is introduced. In this procedure, neither an optimization process is carried out nor is the control structure fixed. But, zeros' location impact over the system dynamical responses is analyzed, showing that a careful selection of the PEC passive elements could both avoid electronic device failures due to large overshoots and improve the dynamical system performance.

The structure of the chapter is organized as follows: in Section 2, both time- and frequency-domain models of the boost DC-DC converter are derived. In Section 3, the boost DC-DC converter is studied in steady state. Section 3 is composed of Sections 3.1, 3.2, 3.3, and 3.4. In Sections 3.1 and 3.2, expressions for $M(D)$ and η are derived including some parasitic losses. In Section 3.3, conditions to operate in CCM or DCM are found. In Section 3.4, both inductor current and capacitor voltage ripple analyses are carried out to find lower boundaries for inductor and capacitor values that fulfill ripple requirements. In Section 4, the value of the passive elements is selected such that operating requirements are fulfilled and system dynamical performance is achieved. Mathematical model is contrasted with a PSIM implementation of the boost DC-DC converter. In Section 5, the widely accepted current mode control (CMC) structure for boost DC-DC converters is designed.

2. Nonlinear dynamical modeling

Figure 1 shows a circuital representation of a typical boost DC-DC converter including its parasitic losses associated to the passive elements. The boost DC-DC

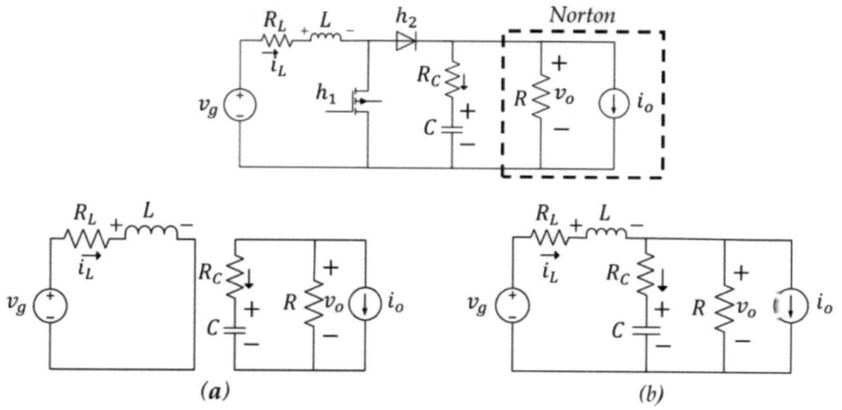

Figure 1.
Circuital scheme of the DC-DC boost converter: configuration (a) $h_1 = 1, h_2 = 0$. Configuration (b) $h_1 = 0, h_2 = 1$.

converter supplies energy to a dominant-current load represented as a Norton equivalent model. Engines and inverters are common dominant-current loads that can be supplied by a boost DC-DC converter. In **Figure 1**, L is inductor, C is capacitor, and R_L and R_C are the parasitic losses for L and C, respectively. R_L and R_C represent all parasitic losses.

The boost DC-DC converter operating in CCM can take two configurations according to the switch position as shown in **Figure 1**. First (**Figure 1(a)**) and second (**Figure 1(b)**) configurations correspond to switch $H = \{h_1, h_2\}$ being turned on $h_1 = 1, h_2 = 0$ and turned off $h_1 = 0, h_2 = 1$, respectively. Therefore, the switching function u can be defined as follows: $u = h_1$ or $u = 1 - h_2$.

State variables are inductor current i_L and capacitor voltage v_C which represent the energy variation in the system. The system inputs are u, DC input voltage source v_g, and current source i_o. Variations of i_o are useful to represent system current perturbations. The system outputs are output voltage v_o and i_L. The corresponding dynamical model of the system in **Figure 1** is given by Eqs. (1) (2), where $\alpha_C = R_C/R$ and $\phi_C = R_C/1 + \alpha_C$.

$$L\frac{di_L}{dt} = v_g - \left(R_L + \phi_C(1-u)^2\right)i_L + \left(\frac{\phi_C}{R} - 1\right)(1-u)v_C + \phi_C(1-u)i_o \tag{1}$$

$$C\frac{dv_C}{dt} = \left(\frac{1}{1+\alpha_C}\right)\left((1-u)i_L - \frac{v_C}{R} - i_o\right) \tag{2}$$

In this chapter, the widely accepted PI-based CMC structure for the boost DC-DC converter is adopted [15, 16]. PI controllers' tuning requires a frequency-domain model. From Eqs. (1) and (2), it is possible to obtain a linear state-space model of the boost DC-DC converter. Next, the frequency-domain model is obtained by means of the realization given by Eq. (3).

$$G(s) = \frac{1}{\det(sI - A)}C[\operatorname{adj}(sI - A)]^T B + D \tag{3}$$

The linear state-space model for the boost DC-DC converter is given by Eqs. (4) and (5), where $x = [i_L, v_C]^T$, $u = [d, v_g, i_o]^T$, and $y = [i_L, v_o]^T$. I_L and V_C, D and I_o are states and inputs in their rated values, respectively. $d = \langle u \rangle_o$ (average value of u) is the duty ratio, a continuous variable, and $d\in[0,1]$. In this chapter, d is used as the input control, while D is d in the operation point.

$$\dot{x} = Ax + Bu \tag{4}$$

$$y = Cx + Du \tag{5}$$

where,

$$A = \begin{bmatrix} -\dfrac{(R_L+\phi_C)(1-D)^2}{L} & \dfrac{\left(\dfrac{\phi_C}{R}-1\right)(1-D)}{L} \\ \dfrac{(1-D)}{(1+\alpha_C)C} & -\dfrac{1}{RC(1+\alpha_C)} \end{bmatrix} \tag{6}$$

$$B = \begin{bmatrix} \dfrac{\left(2\phi_C I_L(1-D) - V_C\left(\dfrac{\phi_C}{R}-1\right) - \phi_C I_o\right)}{L} & \dfrac{1}{L} & \dfrac{\phi_C(1-D)}{L} \\ -\dfrac{I_L}{(1+\alpha_C)C} & 0 & -\dfrac{1}{(1+\alpha_C)C} \end{bmatrix} \tag{7}$$

$$C = \begin{bmatrix} 1 & 0 \\ \phi_C(1-D) & 1-\dfrac{\phi_C}{R} \end{bmatrix} \tag{8}$$

$$D = \begin{bmatrix} 0 & 0 & 0 \\ -\phi_C I_L & 0 & -\phi_C \end{bmatrix} \tag{9}$$

The transfer functions given by Eqs. (10)–(15) are obtained by applying the realization given by Eq. (3), where $G_{i_{Ld}}(s) = I_L(s)/D(s)$, $G_{i_L v_g}(s) = I_L(s)/V_g(s)$, $G_{i_L i_O}(s) = I_L(s)/I_O(s)$, $G_{v_o d}(s) = V_o(s)/D(s)$, $G_{v_o v_g}(s) = V_o(s)/V_g(s)$, and $G_{v_o i_O}(s) = V_o(s)/I_O(s)$.

$$G_{i_{Ld}} = \frac{\left\{\begin{array}{l} \left(\dfrac{1}{R}\right)(RC(1+\alpha_C)(2R\phi_C I_L(1-D) - V_C(\phi_C - R) - R\phi_C I_o)s \\ +(\phi_C+R)(1-D)RI_L - V_C(\phi_C-R) - R\phi_C I_o) \end{array}\right\}}{(RLC(1+\alpha_C))s^2 + \left[L + RC(R_L+\phi_C)(1+\alpha_C)(1-D)^2\right]s + (R_L+R)(1-D)^2} \tag{10}$$

$$G_{i_L v_g} = \frac{RC(1+\alpha_C)s+1}{(RLC(1+\alpha_C))s^2 + \left[L + RC(R_L+\phi_C)(1+\alpha_C)(1-D)^2\right]s + +(R_L+R)(1-D)^2} \tag{11}$$

$$G_{i_L i_O} = \frac{\phi_C RC(1-D)(1+\alpha_C)s + R(1-D)}{(RLC(1+\alpha_C))s^2 + \left[L + RC(R_L+\phi_C)(1+\alpha_C)(1-D)^2\right]s + +(R_L+R)(1-D)^2} \tag{12}$$

$$G_{v_o d} = \frac{\left\{\begin{array}{l} [C\phi_C(1-D)(1+\alpha_C)(2R\phi_C I_L(1-D) - V_C(\phi_C-R) - R\phi_C I_o) + LI_L(\phi_C-R)]s \\ -I_L(\phi_C-R)(1-D)^2(2\phi_C+R_L) + (1-D)(2R\phi_C I_L(1-D) - V_C(\phi_C-R) - R\phi_C I_o) \end{array}\right\}}{(RLC(1+\alpha_C))s^2 + \left[L + RC(R_L+\phi_C)(1+\alpha_C)(1-D)^2\right]s + (R_L+R)(1-D)^2} \tag{13}$$

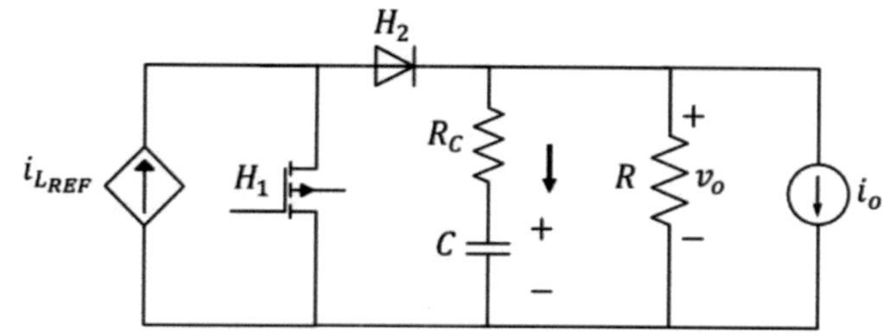

Figure 2.
Equivalent simplified representation of the boost DC-DC converter.

$$G_{v_o v_g} = \frac{R(\phi_C C(1+\alpha_C)s+1)(1-D)}{(RLC(1+\alpha_C))s^2 + \left[L + RC(R_L+\phi_C)(1+\alpha_C)(1-D)^2\right]s + (R_L+R)(1-D)^2} \tag{14}$$

$$G_{v_o i_0} = \frac{R\left(\phi_C(1-D)^2 - \phi_C(1-D) - R_L\right)(\phi_C C(1+\alpha_C)s+1)}{(RLC(1+\alpha_C))s^2 + \left[L + RC(R_L+\phi_C)(1+\alpha_C)(1-D)^2\right]s + (R_L+R)(1-D)^2} \tag{15}$$

Once the current control loop in the CMC structure is closed, the equivalent simplified representation of the boost DC-DC converter shown in **Figure 2** is obtained. Large- and small-signal models of the simplified boost DC-DC converter are given by Eqs. (16) and (17), respectively. The transfer functions of the simplified model are given by Eq. (18). The numerator of Eq. (18) has two components, one for each system input, i.e., i_{REF} and i_o, respectively. In the CMC structure, Eq. (18) is employed to tune the PI controller in the outer control loop, which regulates v_o.

$$C\frac{dv_C}{dt} = \left(\frac{1}{1+\alpha_C}\right)\left(i_{L_{REF}}(1-D) - \frac{v_C}{R} - i_o\right) \tag{16}$$

$$\begin{aligned}[v_C^{\cdot}] &= \left[-\left(\frac{1}{1+\alpha_C}\right)\frac{1}{RC}\right][v_C] + \left[\left(\frac{1}{1+\alpha_C}\right)\frac{(1-D)}{C} \quad -\left(\frac{1}{1+\alpha_C}\right)\frac{1}{C}\right]\begin{bmatrix} i_{L_{REF}} \\ i_o \end{bmatrix} \\ [v_o] &= \left[\left(\frac{1}{1+\alpha_C}\right)\right][v_C] + \left[\left(\frac{1}{1+\alpha_C}\right)R_C(1-D) \quad -\left(\frac{1}{1+\alpha_C}\right)R_C\right]\begin{bmatrix} i_{L_{REF}} \\ i_o \end{bmatrix}\end{aligned} \tag{17}$$

$$G_{v_o}(s) = \frac{\begin{bmatrix} [(1+\alpha_C)RCR_Cs + R + R_C](1-D) \\ -(1+\alpha_C)RCR_Cs + R + R_C \end{bmatrix}}{(1+\alpha_C)[(1+\alpha_C)RCs+1]} \tag{18}$$

3. Steady-state analysis

Once the system model is obtained, the following analysis might be carried out: (1) derivation of the $M(D)$ expression, (2) losses effect and efficiency expression derivation, (3) condition analyses of CCM and DCM, and (4) inductor current Δi_L and capacitor voltage Δv_C ripple analysis. The aim of these analyses is to determine suitable passive elements' (L and C) boundaries which satisfy the design requirements.

3.1 First step: derivation of the equilibrium conversion ratio M(D) expression

Steady-state model allows to derive expressions for average rated values for both v_C and i_L as functions of the system inputs and parameters. The steady-state model is obtained by setting the model given by Eqs. (1) and (2) to zero. Thus, Eqs. (19) and (20) are obtained.

$$I_L = \frac{V_C}{R(1-D)} \tag{19}$$

$$V_g = \left(\frac{R_L + R(1-D)^2}{R(1-D)}\right) V_o \tag{20}$$

From Eqs. (1) and (2), it is found that $V_o = V_C$ in steady state. The expression of $M(D)$ for the boost DC-DC converter is conveniently written using Eq. (21), where $\alpha_L = R_L/R$. $M(D)$ indicates the conversion gain factor in voltage in terms of D, R, and R_L.

$$M(D) = \frac{V_o}{V_g} = \frac{(1-D)}{\alpha_L + (1-D)^2} \tag{21}$$

Note that $M(D)$ does not depend on R_C due to the fact that the capacitor current in the average model is zero, leading to no voltage drop in R_C.

It is important to remark that if $R_L = 0$ in Eq. (21), this expression is in agreement with the ideal boost DC-DC $M(D)$, i.e., $M(D) = 1/(1-D)$. However, the converter reaches an efficiency equal to 100% if $R_L = 0$. Additionally, $M(D)$ tends to ∞ when D tends to 1. The above consideration is not true in a real boost DC-DC converter application and, for this reason, an analysis without including parasitic losses is not convenient.

3.2 Second step: losses effect and efficiency expression derivation

This section shows how parasitic losses affect η in the boost DC-DC converter case. Losses effect and efficiency analyses are carried out in order to find suitable values for R_L and R_C such that the designed PEC fulfills the operating requirements.

The DC transformer correctly represents the relations between DC voltages and currents of the converter. The resulting model can be directly solved to find voltages, currents, losses, and efficiency in the boost DC-DC converter [20].

Eqs. (22) and (23) are obtained from Eqs. (1) and (2). These equations establish that the average value of both i_L and v_C are equal to zero in steady state. **Figure 3** is the representation of Eqs. (22) and (23) as a DC transformer model.

$$0 = V_g - \left(R_L + \phi_C(1-D)^2\right)I_L - \underbrace{\left(\frac{1}{1+\alpha_C}\right)V_C(1-D)}_{V_d} \tag{22}$$

$$0 = -\frac{V_C}{(1+\alpha_C)R} + \underbrace{\left(\frac{1}{1+\alpha_C}\right)I_L(1-D)}_{I_d} \tag{23}$$

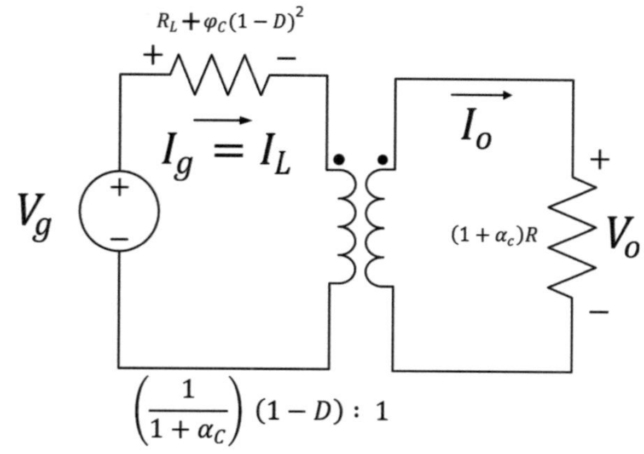

Figure 3.
DC transformer model of the boost DC-DC converter.

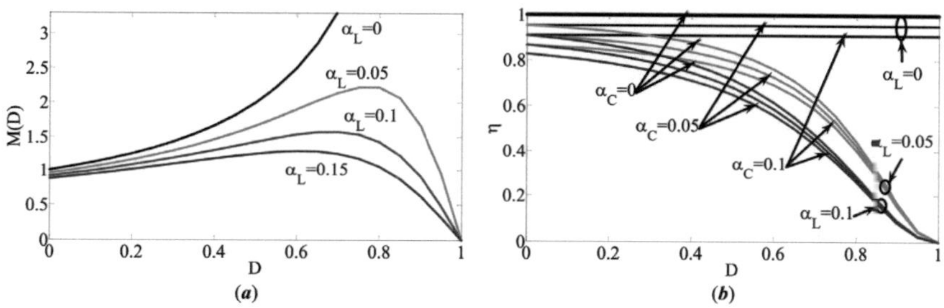

Figure 4.
(a) Conversion ratio M(D) vs. duty cycle D. (b) Efficiency η vs. duty cycle D.

The equivalent circuit model in **Figure 3** allows to compute the converter efficiency η. From **Figure 3**, it is possible to deduce the efficiency expression given by Eq. (24).

$$\eta = \frac{(1-D)^2}{(1+\alpha_C)\left(\alpha_L + (1-D)^2\right)} \tag{24}$$

Simulations of Eqs. (21) and (24) are shown in **Figure 4** for several values of α_C and α_L ratios in order to test how much losses affect both $M(D)$ and η.

Figure 4(a) and **(b)** is shown together to relate $M(D)$ and η. Two different values of D can be selected to reach the same value of $M(D)$. Nevertheless, higher values of D lead to lower efficiency values. For this reason, it is recommended that the converter operates at low values of D as possible.

Figure 4(a) shows how the α_L ratio affects $M(D)$: $\alpha_L = 0$ is the ideal case for the boost DC-DC converter (without losses) and $M(D)$ in the converter has an increasing trend and eventually tends to infinity. When α_L increases (real case, converter with losses), $M(D)$ decreases and the curve has a quadratic trend. It can be observed that the higher α_L value matches to the lower converter conversion ratio $M(D)$.

From **Figure 4(b)**, it is observed that the maximum η value reached by the converter is determined by losses and it is given in $D = 0$ for every $M(D)$ curve. For the studied case, it is the combination of α_C and α_L that determines the maximum value of η, which decreases while D increases, dropping to 0 when D tends to 1.

Hence, the converter should operate as far as possible with low D values. Additionally, an increase of either α_C or α_L causes a decrease in η. Therefore, α_C and α_L should tend to zero to guarantee high converter efficiency. Values of α_C and α_L were clustered in groups of curves. For $\alpha_L = 0$ and $\alpha_C = [0, 0.05, 0.1]$ (above lines group), it is noted that while α_C increases, η slightly decreases. For curves group $\alpha_L = 0.05$ and $\alpha_C = [0, 0.05, 0.1]$ (middle lines group) and for curves group $\alpha_L = 0.1$ and $\alpha_C = [0, 0.05, 0.1]$ (below lines group), something similar occurs—η decreases while α_C increases. However, decreases in η are more notable when α_L increases than when α_C increases. The combined effect of high α_L and α_C leads to a highly inefficient system with high losses.

3.3 Third step: conditions for the converter conduction model

The CCM is suggested since DCM causes a larger voltage ripple in the boost DC-DC converter case [18, 19]. In consequence, the peak inductor current in DCM is higher than in CCM [22]. By [20], the condition for operating in the CCM is $|I_L|>|\Delta i_L|$ and the condition for operating in the DCM is $|I_L|<|\Delta i_L|$. The DCM condition for the boost DC-DC converter is given by Eq. (25), where $T_s = 1/f_{sw}$ and f_{sw} is the converter switching frequency.

$$D(1-D)^2 > \frac{2L}{RT_s} \tag{25}$$

The left side of Eq. (25) is a function that only depends on D. Here, this function is named as $K(D) = D(1-D)^2$. The right side of Eq. (25) is a dimensionless function that depends on L, R, and T_s, which is named in this chapter as $K = 2L/RT_s$. If L and R are taken as the converter parameters and f_{sw} is fixed, K is a constant and represents the converter measure to operate in CCM and DCM [20]. Large values of K lead to CCM. Small values of K lead to the DCM for some values of D. $K(D)$ is a function that represents the boundary between DCM and CCM. Then, the minimum value of K must be at least equal to the maximum value of $K(D)$, i.e., $\max(K(D)) \leq \min(K)$, if it is desired that the converter always operates in CCM, see **Figure 5**. Therefore, if values for R and T_s are given in the system specifications, a condition for the minimum possible value of L that assures CCM operation is given by Eq. (26), with $\max(K(D)) = 0.148$, that is equal to the critical value of K (K_c).

$$L > 0.148\frac{RT_s}{2} \tag{26}$$

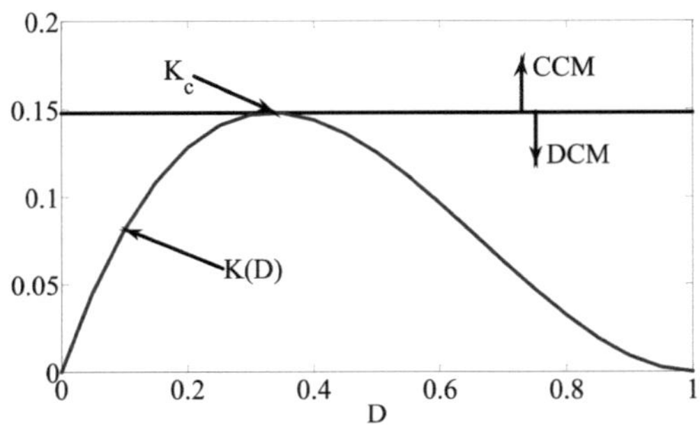

Figure 5.
K(D), K, and conduction mode (CM) conditions.

3.4 Fourth step: inductor current and capacitor voltage ripple analysis

Δi_L and Δv_C analyses are carried out to determine constraint equations for a suitable choice of both L and C values. The carried out analysis in this section is suitable for the boost DC-DC converter operating in CCM. **Figure 6** shows both typical inductor voltage v_L and inductor current i_L linear-ripple approximations. The slope with i_L increasing or decreasing is deduced from the analysis of V_L at each subinterval of time taken into account. Typical values of current inductor ripple Δi_L lie under 10% of the full-load value of I_L [20]. From **Figure 6**, it is seen that i_L begins at the initial value of $i_L(0)$. After time proceeds, i_L increases during the first subinterval (DT_s) and decreases during the second subinterval ($(1-D)T_s$), both with a constant slope. Then, the switch changes back to its initial position at time $t = T_s$ and the process repeats.

As illustrated in **Figure 6(b)**, both current ripple and inductor magnitudes are related through the slope of i_L. The peak inductor current I_{pk} is equal to I_L plus the peak-to-average ripple Δi_L. I_{pk} flows through inductor and semiconductor devices that comprise the switch. The knowledge of I_{pk} is necessary when specifying the rating of the device. The ripple magnitude can be calculated through the knowledge of both the slope of i_L and the length of the first subinterval (DT_s). The i_L linear-ripple approximation is symmetrical to I_L; hence during the first time subinterval, i_L increases by $2\Delta i_L$ (since Δi_L is the peak ripple, the peak-to-peak ripple is $2\Delta i_L$). In consequence, the inductor value L can be chosen from Eq. (27).

$$L = \frac{V_g - \left(\frac{\alpha_L}{(1-D)}\right) V_o}{2\Delta i_L} DT_s \qquad (27)$$

Eq. (27) is a lower boundary for the L value, where L can be chosen such that a maximum Δi_L is attained for the boost DC-DC operating condition.

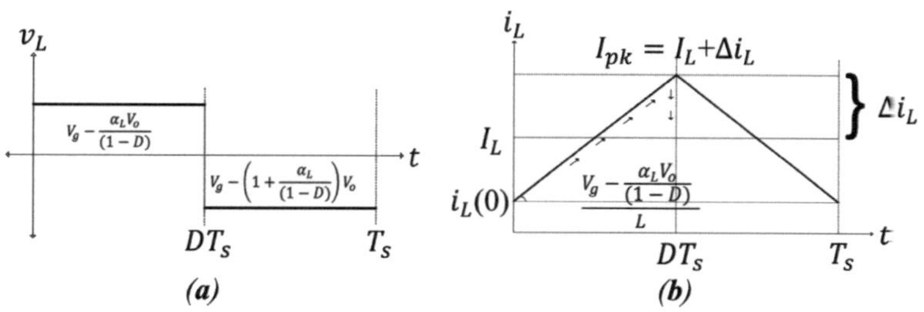

Figure 6.
(a) Typical inductor voltage linear-ripple approximation. (b) Typical current inductor linear-ripple approximation.

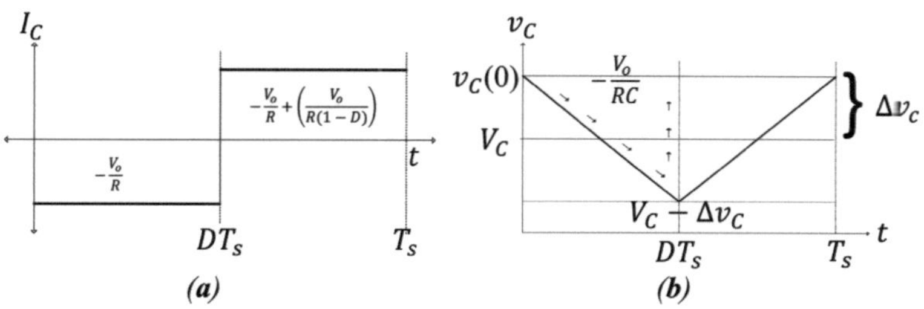

Figure 7.
(a) Typical capacitor current linear-ripple approximation. (b) Typical capacitor voltage linear-ripple approximation.

Likewise, v_C linear-ripple approximation is depicted in **Figure 7(b)**, where a relation between the voltage ripple and the capacitor magnitude is observed. It is seen that v_C begins at the initial value of $v_C(0)$. After time proceeds, v_C decreases during the first subinterval (DT_s) and increases during the second subinterval ($(1-D)T_s$), both with a constant slope. Then, switch changes back to its initial position at time $t = T_s$ and the process repeats itself. The ripple magnitude can be calculated through the knowledge of both the slope of v_C and the length of DT_s. The change in v_C, $-2\Delta v_C$ during DT_s, is equal to the slope multiplied by DT_s. In consequence, Eq. (28) can be used to select the capacitor value of C to obtain a given Δv_C. Eq. (28) is a lower boundary for C value, where C can be chosen such that a maximum Δv_C is attained for the worst boost DC-DC operating condition.

$$C = \frac{V_o}{2R\Delta v_C} DT_s \tag{28}$$

4. Passive elements' value determination

In this section, the boost DC-DC converter operating requirements are specified. Then, the values of the passive elements are determined such that operating requirements are fulfilled and system dynamical performance is achieved. Finally, the mathematical model is contrasted with a PSIM implementation of the boost DC-DC converter.

4.1 System operating requirements

In the boost DC-DC converter application, typical requirements are: input voltage range, output voltage range, output power range, output current range, operating frequency, output ripple, and efficiency. Unless otherwise noted, the continuous operating mode is assumed. The set of operating requirements for the boost DC-DC converter are specified in **Table 1**.

4.2 Zeros' location analysis

The values of the passive elements are selected to fulfill the operating requirements. The main interest is to choose suitable values for inductors and capacitors

Requirements	**Values**		
	Min	**Typ**	**Max**
Input voltage range	30 V	35 V	40 V
Output voltage range	50V	70 V	95 V
Output power range	0 W	100 W	300 W
Output current range	0A	2A	8A (At 50 V)
Operating frequency	100KHz		
Output current ripple	1%	5%	10%
Output voltage ripple	0.1%	0.5%	1%
Steady-state efficiency	90%	95%	98%
Load	25 Ω	50 Ω	100 Ω

Table 1.
Boost DC-DC converter operating requirements.

such that constraints like maximum physical admissible currents and voltages, converter efficiency, and converter CM are satisfied by keeping an acceptable dynamical system performance.

Expressions given by Eqs. (27) and (28) were deduced via steady-state analysis for lower inductor L and capacitor C boundaries, respectively. These expressions are suitable to choose L and C values as functions of Δi_L, Δv_c, states, and inputs in their steady-state value. Additionally, the expression given by Eq. (26) was deduced from the converter CM analysis, which allows to guarantee the boost DC-DC converter CCM operation in all the operating ranges by choosing suitable L and R values.

In PECs, it is desired that $\Delta i_L \leq \max(\Delta i_L)$ and $\Delta v_o \leq \max(\Delta v_o)$ should be assured in the entire operation range. Then, based on the worst condition for Δi_L and Δv_o, L and C lower boundaries can be deduced such that ripple constraints are satisfied. Eqs. (29) and (30) give lower boundaries for L and C, respectively.

$$L \geq \frac{\max(V_g) - \left(\frac{R_L}{\max(R)(1-D)} \min(V_o)\right)}{2\max(\Delta i_L)} DT_s \tag{29}$$

$$C \geq \frac{\max(V_o)}{2\min(R)\max(\Delta v_o)} DT_s \tag{30}$$

Eq. (26) also gives a minimum boundary for L value. Then, Eqs. (26) and (29) must be evaluated and the maximum L value must be selected as the lower boundary.

Eqs. (29) and (26) depend on R_C and R_L through α_C and $M(D)$, respectively. However, from the steady-state analysis, instead of calculating R_C and R_L values, it is suitable to establish α_C and α_L values. α_C and α_L values can be chosen such that the system efficiency is $\eta \geq 90\%$ in the entire operating range.

The maximum boost DC-DC conversion condition corresponds to $\max(M(D))$. Then, $\max(M(D)) = \max(V_o)/\min(V_g)$. According to the operating requirements in Table 1, $\min(V_g) = 30V$ and $\max(V_o) = 95V$, thus the maximum conversion condition in the example here presented is $\max(M(D)) \approx 3.17$. In consequence, loss ratios must be $\alpha_C < 0.05$ and $\alpha_L < 0.05$ when $R = \max(R) = 100\Omega$.

Figure 8 shows $M(D)$ and η curves for $R_L = 150m\Omega$, $R_C = 70m\Omega$, and $R = 25\Omega$, i.e., $\alpha_L = 0.006$ and $\alpha_C = 0.0034$. With these α_L and α_C values, it is assured that $\eta \geq 90\%$ and $M(D) \approx 3.17$. Two points are remarked over both $M(D)$ and η curves for the rated converter conversion condition $M(D) = 2$ and $M(D) \approx 3.17$. From **Figure 8(a)**, it is seen that the converter has sufficient boost capacity to guarantee that for $M(D) \approx 3.17$, the voltage requirement is satisfied. Additionally, from **Figure 8(b)**, it is

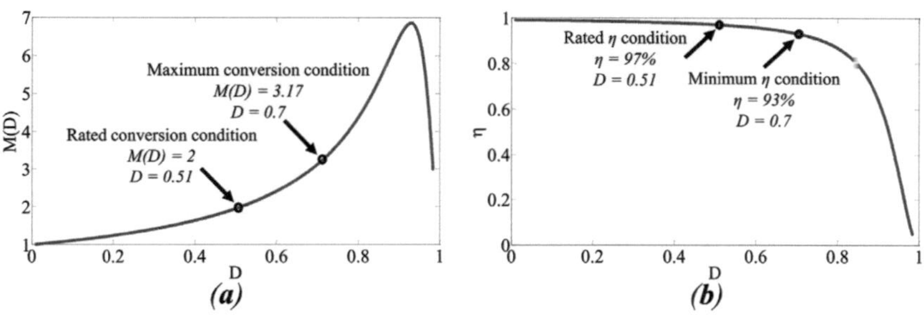

Figure 8.
(a) Conversion ratio $M(D)$. (b) Efficiency η.

seen that for $M(D) = 2$ and $M(D) \approx 3.17$, the converter has 97 and 93% of efficiency, respectively.

From Eq. (29), $\max(I_L) = \max(V_C)/\min(R)(1-D)$. Then, on the one hand, if $\max(\Delta i_L) = 0.1\max(I_L)$, $L \geq 326.34\mu H$ must be selected according to Eq. (29) in order to keep the converter in safe operation [22]. On the other hand, $L \geq 40\mu H$ to always operate in CCM by evaluating Eq. (26). The i_L ripple-based condition is a less restrictive boundary for L than the CCM-based condition. Therefore, $L \geq 326.34\mu H$ is the lower boundary for this element.

If $\max(\Delta v_C) = 0.01\max(V_o)$ in order to keep a converter in safe operation [22], $C \geq 14.120\mu F$ according to Eq. (30).

Minimum L and C values are selected as system parameters. Next, a simulation of the designed boost DC-DC converter is carried out. **Figure 9** shows the step system response for $V_g = 35V$, $V_o = 70V$, $I_o = 0A$, $L = 326.34\mu H$, $C = 14.120\mu F$, $R = 50\Omega$, $\alpha_C = 0.0034$, $\alpha_L = 0.006$, and $f_{sw} = 100kHz$.

From **Figure 9**, it is seen that, with minimum values of L and C, voltage overshoot is $O.S._{G_{vd}} = 57.4718\%$, current overshoot is $O.S._{G_{i_L d}} = 187.2323\%$, and system setting time is $ts = 3.03ms$. From **Figure 9(a)**, it is seen that the peak current value is $33.1686A$, while the steady-state current value is around $11.5477A$. From **Figure 9(b)**, it is seen that the peak voltage value is $214.5027\ V$, while the steady-state voltage value is around $136.2166\ V$.

A designed system with these overshoots needs to oversize its electronic devices such that these devices support both peak voltage and current values without system damage. However, such electronic devices can be expensive and inconvenient. For instance, in this chapter, an analysis of the boost DC-DC converter dynamical characteristics is carried out. This dynamical analysis studies the impact of L and C values over the zeros in transfer functions given by Eqs. (10) and (13), which determine overshoots and system setting time.

In a system as it is known, dynamical response is determined by poles and zeros' location [23]. Zeros are determined by the selected inputs and outputs of the system. Zeros' location is related to some system performance restrictions such as tracking limitations in feedback systems when classical control structures are employed [21, 22]. Moreover, large current or voltage overshoots in converter transient response can cause converter failures. PEC design process could take into account zeros' location due to L and C values such that the right half-plane (RHP) zeros are avoided or their impacts are attenuated. In consequence, the impact of L and C values over the zeros is analyzed to establish a trade-off between their values and the dynamical system response.

In the boost DC-DC converter d is chosen as control input, while v_g and i_o are considered disturbances. Thus, d variations' effect is of primary interest over system output. In consequence, duty ratio-to-voltage-output (G_{vd}) and duty-ratio-to-inductor-current ($G_{i_L d}$) transfer functions are studied. A simulation was carried out

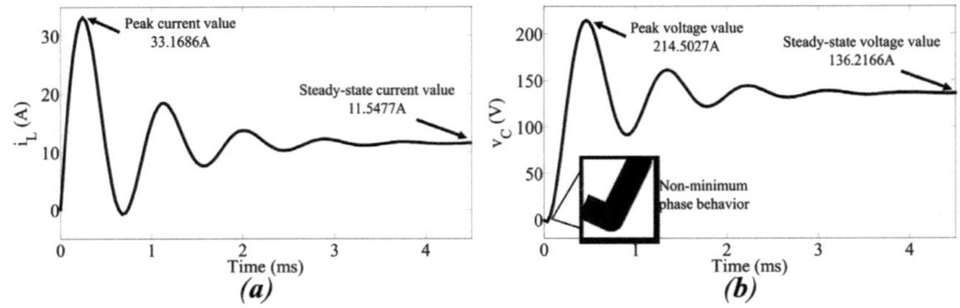

Figure 9.
(a) $G_{i_L d}$ step system response. (b) G_{vd} step system response.

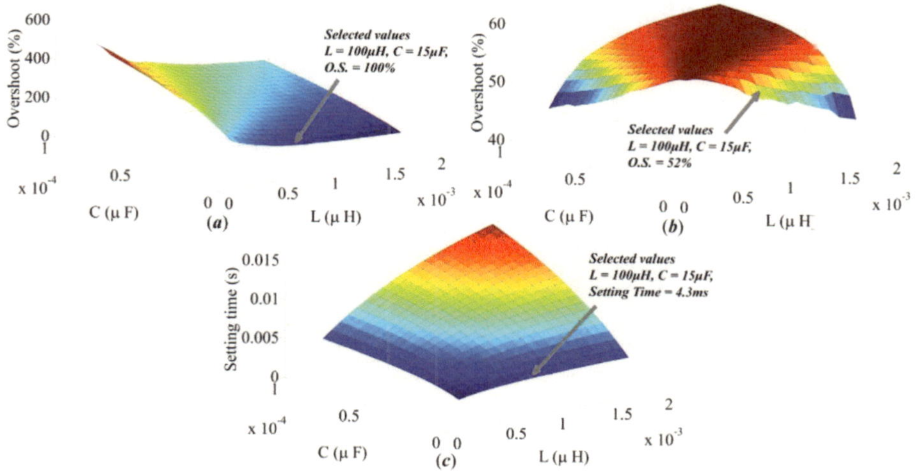

Figure 10.
(a) G_{vd} step system response for varying L and C: overshoot with zeros. (b) G_{i_Ld} step system response for varying L and C: overshoot with zeros. (c) Step system response varying L and C: setting time.

to evaluate the effects of large values for both L and C. **Figure 10** shows G_{vd} and G_{i_Ld} overshoots and setting time for $L\in[326.34\mu H, 2000\mu H]$ and $C\in[14.12\mu F, 100\mu F]$.

From **Figure 10**, it is seen that the minimum possible value of C causes maximum overshoot in v_o; while a minimum possible value of L causes maximum overshoot in i_L. Moreover, minimum C and L values give minimum system setting time.

In contrast, large values of C cause high overshoot of i_L; while large values of L cause high system setting time. In consequence, two additional design requirements are given in order to establish maximum possible values for L and C such that system overshoots and setting time are suitable: (a) maximum duty-ratio-to-output-voltage overshoot $\max\left(O.S._{G_{vd}}\right)$ and (b) maximum duty-ratio-to-inductor-current overshoot $\max\left(O.S._{G_{i_Ld}}\right)$.

From **Figure 10**, it is seen that the system dynamical response cannot be modified if the values of both L and C are simultaneously increased. Meanwhile, if either L or C values are increased, both $O.S._{G_{vd}}$ and $O.S._{G_{i_Ld}}$ decrease. Nevertheless, larger values of L have a major impact than larger values of C.

$L = 1mH$ and $C = 15\mu F$ are selected by results shown in **Figure 10** since with these values $O.S._{G_{i_Ld}} \approx 105\%$ and $O.S._{G_{vd}} \approx 53\%$, i.e., $O.S._{G_{i_Ld}}$ is approximately reduced to 82% and $O.S._{G_{vd}}$ is approximately reduced to 4%. Furthermore, $ts = 4.32ms$, i.e., the system setting time is only increased by $1.3ms$. Thus, these L and C values establish a trade-off between system overshoots and performance. It is remarked that selected L and C values are commercially available.

4.3 System frequency response verification

Frequency response of both the mathematical model and a PSIM circuital implementation are contrasted in order to validate the dynamical model of the designed boost DC-DC converter via simulation. The boost DC-DC converter was parameterized with $L = 1mH$, $C = 15\mu F$, $V_g = 35V$, $V_o = 70V$, $I_o = 0A$, $\alpha_C = 0.0034$, $\alpha_L = 0.006$, $R = 50\Omega$, and $f_{sw} = 100kHz$. In consequence, $I_L = 2.8812A$ and $D = 0.5141$ in the equilibrium point.

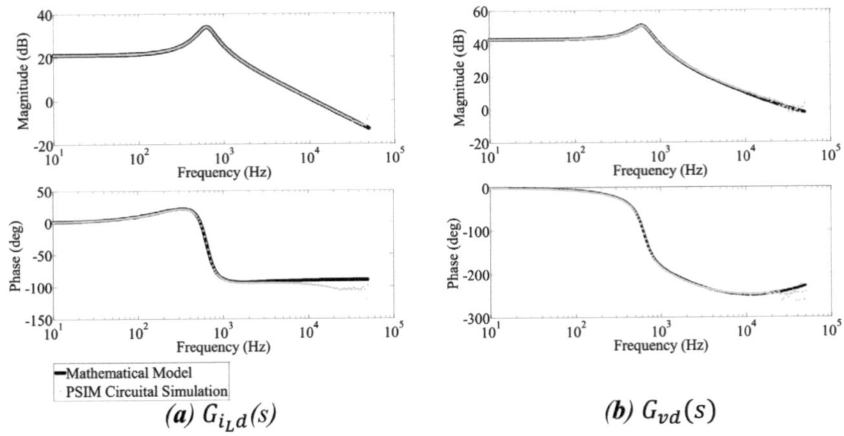

Figure 11.
(a) $G_{i_L d}$ Bode diagram. (b) G_{vd} Bode diagram.

Figure 11 presents the boost DC-DC converter Bode diagrams of the PSIM circuital implementation and the mathematical model given by Eqs. (10)–(15). The frequency response of the PSIM circuital implementation matches with the mathematical model. Then, the PSIM circuital implementation is satisfactorily reproduced by the mathematical model.

5. Control structure design

Nonminimum phase behavior is a well-known result derived from the boost DC-DC converter study [24]. To avoid this system behavior, a CMC structure has been proposed [18, 24]. Nonminimum phase behavior is avoided with this control structure since both $G_{i_L d}$ and inductor-current-to-output-voltage $G_{v_o i_{L_{REF}}}$ transfer functions have a minimum phase behavior.

The converter control design is focused on imposing a desired low-frequency behavior on the system. Here, a CMC structure for the boost DC-DC converter is designed. The aim is to tune PI controllers such that the control objective is achieved. **Figure 12** shows the CMC structure for the DC-DC boost converter. As it is seen in **Figure 12**, the CMC structure employs two PI controllers: first one for i_L control and second one for v_o regulation. These PI controllers are arranged in master-slave form; where i_L control loop is the inner loop and v_o control loop is the outer loop. This master-slave arrangement allows v_o regulation while preserving i_L within specified safety limits.

In the boost DC-DC converter which operates in a switch-mode power supply and feeds a certain variable load, the d needs adjustments in order to ensure a constant v_o for the entire operating range (voltage regulation). Besides, against any system disturbance (v_g and i_o random changes), the d value should be adjusted to drive the system back to the operating point. The PI controller in the outer loop provides the set-point of the inner loop, which acts as the control input of the outer loop. The proportional and integral (PI) controller in the inner loop generates a continuous signal for d, which by means of a pulse width modulation (PWM) is applied to the power switching gate.

5.1 Controller tuning

Controller's tuning task begins with the set of design specifications. The goal of the boost DC-DC converter controller is to maintain v_o within 2% of its rated value

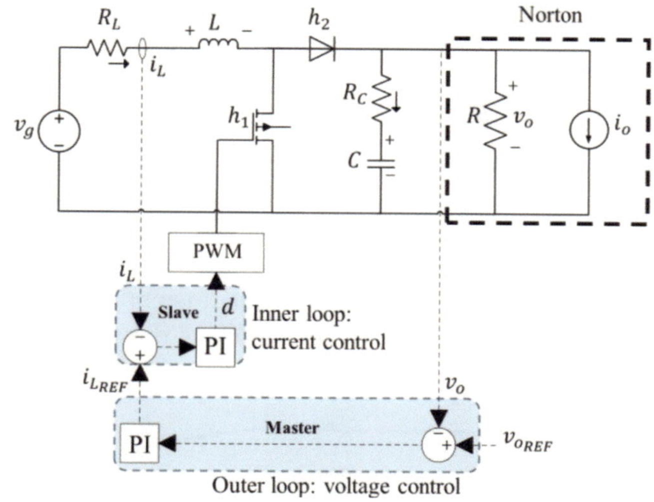

Figure 12.
Boost DC-DC CMC structure scheme.

(i.e., 68.6V–71.4V) in response to random changes (disturbances) in both v_g and i_o. Also, the controller should be able to drive v_o within the tolerance for v_g variations over a range from 30 to 40V.

The inner loop control bandwidth must be 20kHz or less due to the fact that f_{sw} is equal to 100kHz, and the outer loop control bandwidth must be smaller than 1/5 of the inner loop control bandwidth [25], i.e., smaller than 5kHz. Additionally, a robustness index of M_s<2 is desired to establish a trade-off between control performance and robustness [26].

A PI controller was tuned by acting directly on d to track the inductor current reference $i_{L_{REF}}$ since $G_{i_L d}$ exhibits a minimum phase behavior. The inductor current PI controller was tuned by means of the root-locus technique, adopting the following design specifications: damping factor ζ equal to 0.707 and a 20kHz closed-loop bandwidth. The tuned PI controller transfer function $G_{C_{i_L}}(s)$ is given by Eq. (31). These PI controller design specifications ensure: (a) Zero steady-state error and a satisfactory reference tracking for frequencies below 20kHz; this is observed on transfer function $T_{i_L i_{L_{REF}}}$ in **Figure 13**. (b) Effective disturbance rejection for both input voltage v_g and current source i_o variations, which are observed on transfer functions $T_{i_L v_g}$ and $T_{i_L i_o}$ in **Figure 13**, respectively. (c) A closed-loop robustness $M_s = 1.2$.

$$G_{C_{i_L}}(s) = \frac{1.27s + 55218}{s} \quad (31)$$

A PI controller was tuned to regulate v_o since the $G_{v_o i_{L_{REF}}}(s)$ transfer function given by the Eq. (18) exhibits a minimum phase behavior. This PI controller provides the set-point of the inner control loop. The PI controller of the outer control loop was tuned by means of the root-locus technique considering a damping factor ζ equal to 0.707 and a 5kHz closed-loop bandwidth. The tuned PI controller transfer function $G_{C_{v_o}}(s)$ is given by Eq. (32). These PI controller design specifications ensure: (a) Zero steady-state error observed on transfer function $T_{v_o v_{o_{REF}}}$ in **Figure 14**. (b) Effective disturbance rejection for the current source i_o variations, which are observed on transfer function $T_{v_o i_o}$ in **Figure 14**. (c) A closed-loop robustness $M_s = 1.2$.

$$G_{C_{v_o}}(s) = \frac{0.07994s + 235.1}{s} \tag{32}$$

5.2 Closed-loop system performance verification

The designed boost DC-DC converter with its control structure was implemented in PSIM to assess the closed-loop system robustness. Three cases were proposed to evaluate the control structure robustness against most common disturbances. (i) An experiment that simulates a change of ±35% around the nominal value of the load was carried out. Next, (ii) an experiment that simulates a change of ±15% around the nominal value of the input voltage was carried out. Finally,

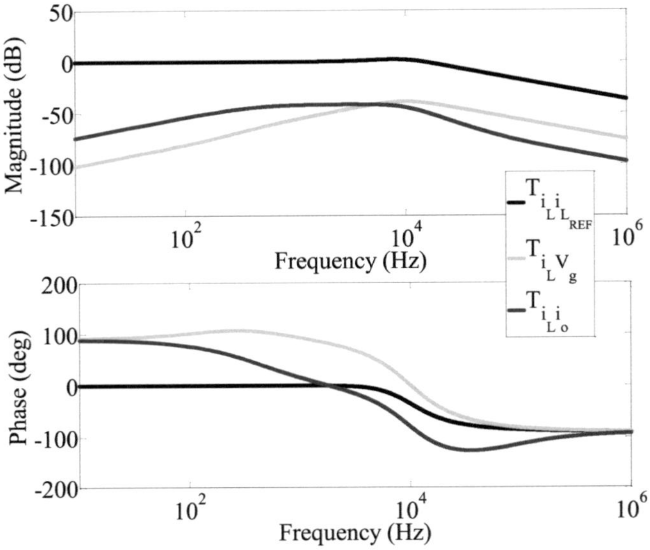

Figure 13.
Inner current control loop transfer functions.

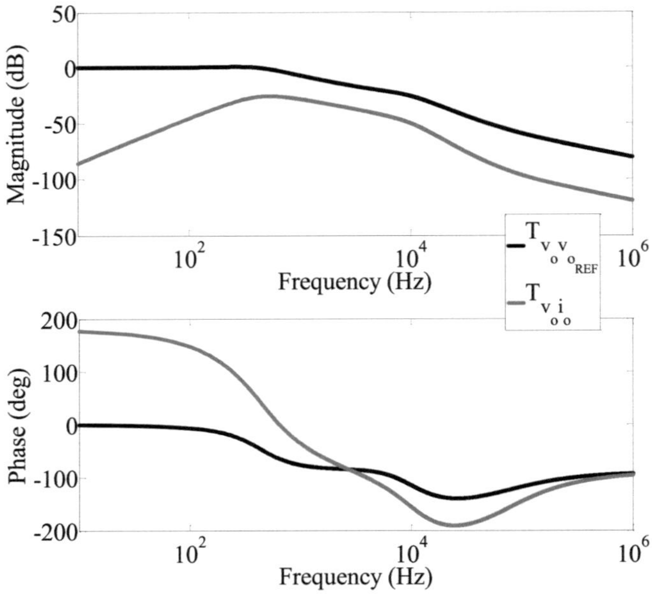

Figure 14.
Outer output voltage control loop transfer functions.

(iii) an experiment that simulates a combined change of ±35 and ±30% around the nominal values of the load and the input voltage, respectively, was carried out. **Figure 15** shows the dynamical system response against the perturbations mentioned above. From **Figure 15**, it is seen that the system stability is not affected by any of the simulated perturbations, which means that the control structure is robust against the system perturbations from both the load and the input voltage up to 35%.

Figure 15(a) shows the closed-loop behavior at unit step changes of i_o around the operating point corresponding to the full load. Two i_o unit step changes were applied to evaluate the control structure performance. The first step change was applied at $t = 10ms$ for $10ms$, then the current source returns to its rated value $i_o = 0A$. The second unit step change was applied at $t = 30ms$ for $10ms$, then the current source returns to its rated value $i_o = 0A$. In **Figure 15(a)**, a satisfactory tracking of $i_{L_{REF}}$ and regulation of v_o to reject load disturbances depicted as changes in i_o is observed.

Figure 15(b) shows the closed-loop behavior at unit step changes of v_g. Two v_g unit step changes were applied to evaluate the control structure capabilities to regulate v_o and to evaluate the capabilities of the designed boost DC-DC converter. The first unit step change was applied at $t = 10ms$ for $10ms$. This first unit step change was equal to $v_g = -5V$, i.e., the final value of the input voltage was $v_g = 30V$ that corresponds with its lower boundary. The second unit step change was applied at $t = 30ms$ for $10ms$. This second unit step change was equal to $v_g = +5V$, i.e., the final value of the input voltage was $v_g = 40V$ that corresponds to its upper boundary. In **Figure 15(b)** a satisfactory reference tracking of $i_{L_{REF}}$ and control regulation of v_o to changes in v_g is observed. It is important to remark that under the worst condition for v_g, the boost DC-DC converter was able to keep v_o in its rated value.

Finally, **Figure 15(c)** shows the closed-loop behavior at random unit step changes of both i_o and v_g. These unit step changes were applied such that the designed control structure performance could be evaluated against any random disturbance. From **Figure 15(c)**, it is possible to see that the designed control structure has a satisfactory performance against multiple disturbances within specified design requirements for the boost DC-DC converter in **Table 1**.

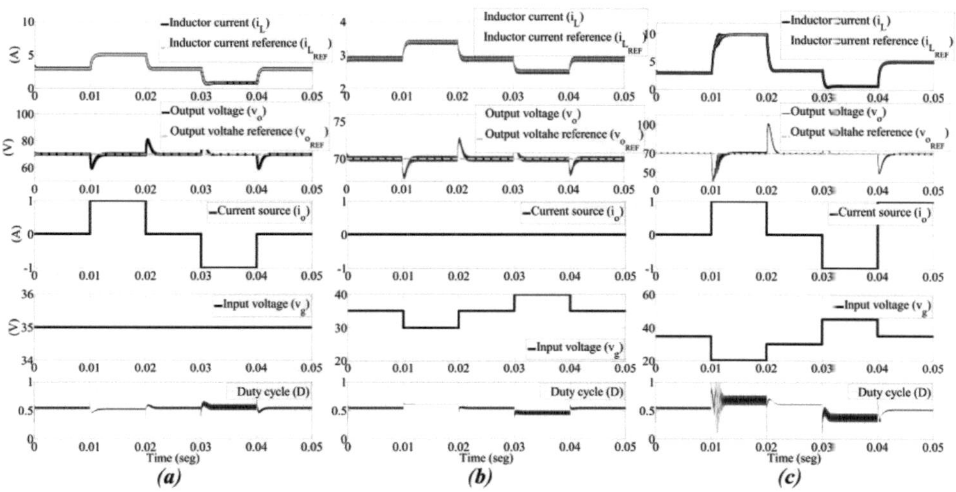

Figure 15.
Closed-loop behavior at unit steps system disturbances.

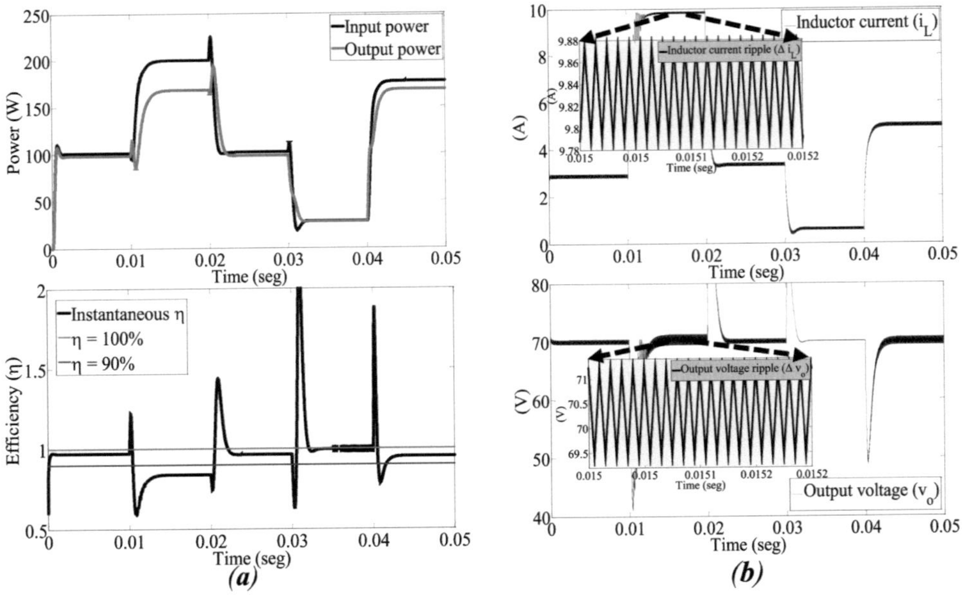

Figure 16.
(a) Instantaneous Power verification. (b) Ripples verification.

In order to carry out system operation requirements verification, case (c) of **Figure 15** is taken into account. **Figure 16** shows: (a) P_{in}, P_{out} and (b) i_L and v_o, when case (c) of **Figure 15** is considered.

From **Figure 16(a)**, it is seen that P_{out} does not exceed the maximum admissible output power in steady state and is always lower than P_{in}. (b). **Figure 16(b)** shows i_L and v_o. A zoom was made for the worst simulated system condition. From **Figure 16(b)**, it is seen that even in the worst i_L and v_o condition, Δi_L and Δv_o are below 1%. Accordingly, the designed boost DC-DC converter satisfies both Δi_L and Δv_o conditions.

In conclusion, **Figure 16** shows that the boost DC-DC converter system operating requirements given in Table 1 are successfully satisfied.

6. Conclusions

In this chapter, a procedure to easily design and control PECs was proposed and zeros' location impact over the system dynamical responses was analyzed, showing that a careful selection of the PEC passive elements could both avoid electronic device failure due to large overshoots and improve the dynamical system performance. Parasitic losses R_L and R_C were included in order to have a more realistic approach to the system. The presented procedure was composed of:

a. The nonlinear dynamical system modeling approach to obtain a mathematical tool and evaluate the system performance. The obtained dynamical model was suitable to describe the dynamical behavior of the system and to derive the steady-state model.

b. The steady-state analysis that allowed to find suitable constraints for passive elements' values. The steady-state analysis was composed of: (a) $M(D)$ expression derivation, (b) losses effect analysis and η expression derivation, (c) conditions for analysis of CCM and DCM, and (d) Δi_L and Δv_C analyses.

c. The passive elements' value determination based on the system zeros' location. A zero-based analysis allowed to choose the passive elements' values such that a trade-off between operating requirements and system transient response were achieved. This analysis reduced the system outputs' overshoot alleviating electronic devices' stress and improving the system's performance.

d. The proposal of a model-based control structure; particularly, a CMC structure based on PI controllers for automatic converter control was implemented in the boost DC-DC converter, although a control structure does not need to be fixed in this procedure.

e. The procedure was applied to a boost DC-DC converter application taking into account the parasitic losses associated with its passive elements, which allows to investigate the details of its performance, operation, and behavior. It was possible to design a boost DC-DC converter that fulfills all the operating requirements in the entire operating range, even if bounded disturbances appear. Design was based on the nonlinear dynamical model and steady-state analysis. The CMC structure was implemented for the designed boost DC-DC converter. PI controllers were tuned by means of root-locus controller design method. The boost DC-DC converter was implemented in PSIM where system operating requirements, closed-loop performance, and robustness were successfully verified.

Acknowledgements

This work was partially supported by COLCIENCIAS (Fondo Nacional de Financiamiento para la Ciencia, la Tecnología y la Innovación Francisco José de Caldas) with the doctoral scholarship 727-2015. The authors would also like to thank "estrategia sostenibilidad UdeA".

Conflict of interest

The authors declare no conflict of interest.

Author contributions

Jorge H. Urrea-Quintero carried out most of the work presented here. Nicolás Muñoz-Galeano was the advisor of this work, and Lina M. Gómez had a relevant contribution with her extensive knowledge about systems theory.

Author details

Jorge-Humberto Urrea-Quintero[1*], Nicolás Muñoz-Galeano[1] and
Lina-María Gómez-Echavarría[2]

1 Faculty of Engineering, Universidad de Antioquia, Medellín, Colombia

2 Department of Process and Energy, Universidad Nacional de Colombia,
Sede Medellín, Colombia

*Address all correspondence to: humberto.urrea@udea.edu.co

References

[1] Liu J, Hu J, Xu L. Dynamic modeling and analysis of z source converter—Derivation of AC small signal model and design-oriented analysis. IEEE Transactions on Power Electronics. 2007;**22**(5):1786-1796

[2] Arango E, Ramos-Paja CA, Calvente J, Giral R, Serna-Garces SI. Asymmetrical interleaved DC/DC switching converters for photovoltaic and fuel cell applications—Part 2: Control-oriented models. Energies. 2013;**6**(10):5570

[3] Liang TJ, Tseng KC. Analysis of integrated boost-flyback step-up converter. IEE Proceedings—Electric Power Applications. 2005;**152**(2):217-225

[4] Davoudi A, Jatskevich J, Chapman PL, Bidram A. Multi-resolution modeling of power electronics circuits using model-order reduction techniques. IEEE Transactions on Circuits and Systems I. 2013;**60**(3):80-823

[5] Beldjajev V, Roasto I. Efficiency and voltage characteristics of the bi-directional current doubler rectifier. Przeglad Elektrotechniczny. 2012;**88**(8):124-129

[6] Galigekere VP, Kazimierczuk MK. Analysis of PWM Z-source DC-DC converter in CCM for steady state. IEEE Transactions on Circuits and Systems I. 2012;**59**(4):854-863

[7] Geyer T, Papafotiou G, Frasca R, Morari M. Constrained optimal control of the step-down DC-DC converter. IEEE Transactions on Power Electronics. 2008;**23**(5):2454-2464

[8] Vlad C, Rodriguez-Ayerbe P, Godoy E, Lefranc P. Advanced control laws of DC-DC converters based on piecewise affine modelling. Application to a stepdown converter. IET Power Electronics. 2014;**7**(6):1482-1498

[9] Chiniforoosh S et al. Definitions and applications of dynamic average models for analysis of power systems. IEEE Transactions on Power Delivery. 2010;**25**(4):2655-2669

[10] Trejos A, Gonzalez D, Ramos-Paja CA. Modeling of step-up grid-connected photovoltaic systems for control purposes. Energies. 2012;**5**(6):1900

[11] Kapat S, Krein PT. Formulation of PID control for DC-DC converters based on capacitor current: A geometric context. IEEE Transactions on Power Electronics. 2012;**27**(3):1424-1432

[12] Kabalo M, Paire D, Blunier B, Bouquain D, Godoy Simões M, Miraoui A. Experimental evaluation of four-phase floating interleaved boost converter design and control for fuel cell applications. IET Power Electronics. 2013;**6**(2):215-226

[13] Mapurunga Caracas JV, De Carvalho Farias G, Moreira Teixeira LF, De Souza Ribeiro LA. Implementation of a high-efficiency, high-lifetime, and low-cost converter for an autonomous photovoltaic water pumping system. IEEE Transactions on Industry Applications. 2014;**50**(1):631-641

[14] Evzelman M, Ben-Yaakov S. Simulation of hybrid converters by average models. IEEE Transactions on Industry Applications. 2014;**50**(2):1106-1113

[15] Gezgin C, Heck BS, Bass RM. Integrated design of power stage and controller for switching power supplies. In: IEEE Workshop on Computers in Power Electronics; 1996; pp. 36-44

[16] Pomar M, Gutierrez G, de Prada Moraga C, and Normey J E. Rico. Integrated design and control applied to a buck boost converter. In: European Control Conference; 2007, pp. 948-954

[17] Mariethoz S, Almer S, Morari M. Integrated control and circuit design; an optimization approach applied to the buck converter. In: American Control Conference (ACC); 2010, pp. 3305-3310

[18] Alvarez-Ramirez J, Espinosa-Pérez G, Noriega-Pineda D. Current-mode control of DC-DC power converters: A backstepping approach. International Journal of Robust and Nonlinear Control. 2003;**13**(5):421-442

[19] Qiu Y, Liu H, Chen X. Digital average current-mode control of pwm dc-dc converters without current sensors. IEEE Transactions on Industrial Electronics. 2010;**57**(5):1670-1677

[20] Erickson RW, Maksimovic D. Fundamentals of Power Electronics. Springer Science & Business Media; 2007

[21] Ping L, Xin M, Bo Z, Zhao-ji L. Analysis of the stability and ripple of PSM converter in DCM by EB model. In: International Conference on Communications, Circuits and Systems; 2007. pp. 1240-1243

[22] Liu S-L, Liu J, Mao H, Zhang Y. Analysis of operating modes and output voltageripple of boost DC-DC convertersand its design considerations. IEEE Transactions on Power Electronics. 2008;**23**(4):1813-1821

[23] Hoagg JB, Bernstein DS. Nonminimum-phase zeros: Much to do about nothing—classical control revisited. Part II. IEEE Control Systems Society. 2007;**27**(3):45-57

[24] Chen Z, Gao W, Hu J, Ye X. Closed-loop analysis and cascade control of a nonminimum phase boost converter. IEEE Transactions on Power Electronics. 2011;**26**(4):1237-1252

[25] Louganski KP, Lai J-S. Current phase lead compensation in single-phase pfc boost converters with a reduced switching frequency to line frequency ratio. IEEE Transactions on Power Electronics. 2007;**22**(1):113-119

[26] Vilanova R, Alfaro VM. Control PID robusto: Una visión panorámica. Revista Iberoamericana de Automatica e Informatica Industrial (RIAI). 2011; **8**(3):141-158

Chapter 8

Energy Efficient Speed Control of Interior Permanent Magnet Synchronous Motor

Olga Tolochko

Abstract

In this chapter, methods for the structural realization of a speed control system for the interior permanent magnet synchronous motor (IPMSM) using the "maximum torque per ampere" (MTA) and "maximum torque per volt" (MTV) optimal control strategies are considered. In the system in constant torque region, is a technique for adapting the speed controller to the presence of the reactive motor torque component, which improves the quality of the transient processes, is proposed. It is also recommended to approximate the dependence of the flux-forming current component on the motor torque by the "dead zone" nonlinearity, which will simplify the optimal control algorithm and avoid solving the fourth-degree algebraic equation in real time. For the speed control with field weakening technique, a novel system is recommended. In this system, the control algorithms are switched by the variable of the direct stator current component constraint generated in accordance with the MTA law: the upper limit is calculated in accordance with the "field weakening control" (FWC) strategy, and the lower limit in accordance with the MTV strategy. The steady-state stator voltage constraint is implemented through the variable quadrature stator current component limitation. The effectiveness of the proposed solutions is confirmed by the simulation results.

Keywords: interior permanent magnet synchronous motor, optimal control, maximum torque per ampere, maximum torque per volt, field weakening control, stator voltage constraint, simulation

1. Introduction

Currently, more attention is being paid to improving the energy efficiency of managing electromechanical plants. The solution for this problem is of particular importance for electric drive systems with autonomous power sources, in particular for electric vehicles, allowing them to increase their mileage between recharges. One of the motors widely used in electric vehicles is the permanent magnet synchronous motor (PMSM).

Depending on the magnets' location in the rotor, PMSMs are divided into motors with surface-mounted magnets (SPMSM—surface permanent magnet synchronous machine) and interior-mounted magnets (IPMSM—interior PMSM). The surface allocation of the magnets prevents the engine from operating at high speed. With the internal allocation of the magnets, the mechanical strength of the rotor increases, and this defect is eliminated. SPMSM has a symmetrical magnetic system, since the magnetic permeability of air and permanent magnets is practically the same. The electromagnetic system in IPMSM is asymmetric. Hence, IPMSM is a salient pole motor. This leads to the occurrence, along with the active component of the torque, of an additional reactive component, through which it is possible to obtain larger power/weight, torque/current, and torque/voltage ratios.

For IPMSM, optimal control strategies have been developed [1–5]. They increase the energy efficiency of electric drives in steady-state conditions by forming the relationship between the stator current orthogonal components, corresponding to the chosen optimality criterion.

With IPMSM speed control, three ranges are possible, each with different control algorithms. The greatest difficulty in the practical implementation of such systems is the organization of control switching from one algorithm to another while maintaining stator current and voltage constraints.

There are different approaches to improving the energy efficiency of the studied electric drive systems. First of all, they differ in the kind of losses that are subject to minimization.

Losses in IPMSM frequency systems consist of losses in windings (copper losses), losses in magnetic conductor (iron losses), losses from higher harmonics in windings and grooves (stray losses), switching losses in the frequency converter, and mechanical losses [6, 7]. In turn, the losses in steel consist of losses for reversal of magnetization (hysteresis losses) and losses from eddy currents, and mechanical losses due to friction losses and losses from air or liquid resistance. The commutation, mechanical, and hysteresis losses are determined by approximate empirical formulas. Therefore, to their analysis, many papers [6–11] are devoted. There are also methods for determining them experimentally [12, 13].

Optimization methods have also different input and output signals in minimizing expressions. For systems of torque and speed control, the most logical is the use of the dependences of the direct and quadrature stator current components on the electromagnetic torque. However, these dependencies are most complex. For example, many of them have the form of equations of the fourth degree, which either have to be solved by iterative methods in real time, or approximated by simpler equations by the least squares method, or represented in the form of precomputed look-up tables (LUTs), the search in which it is performed by interpolation methods. Therefore, optimization equations often are found in the form of dependencies between the components of the stator currents or in the form of the amplitude-phase trajectory of the stator current. In both cases, the quality of transient processes deteriorates.

Minimization of copper losses is provided by the "maximum torque per ampere" (MTA) management strategy, and the "maximum torque per volt" (MTV) strategy minimizes steel losses from eddy currents. These strategies use variable speed ranges in different ranges: MTA for under the rated speed and MTV for over the rated speed.

In [14–16], the static characteristics of the drive using the MTA strategy were analyzed without and with taking into account the saturation of the steel and the demagnetization of the permanent magnets. Based on the analysis performed, a control system is synthesized, in which the speed controller generates a reference to the amplitude of the stator current, and the phase angle is determined from the previously calculated LUT.

It was shown in [17] that in the flux-weakening constant power region, the stator current sometimes becomes uncontrollable in transients because of the current regulator saturation. To eliminate this disadvantage, the authors proposed to set the reference torque of the motor at the output of the speed controller and use another LUT to calculate the reference peak stator current.

In [18], a comparison of the MTA strategy with such additional control methods of energy efficiency increasing as "constant torque angle" (CTA), "unity power factor" (UPF), "constant mutual flux linkage" (CMFL) and "angle control of air gap flux current phasor" (ACAGF). It is shown that UPF control yields a comparatively low voltage requirement but very low torque/current ratio. On comparing UPF with CMFL control, it should be noted that the voltage requirement for CMFLC is next to UPFC but can produce much higher torque/current ratio, which is quite a bit smaller than using the MTA method.

In [19], an equation of the fourth-order polynomial about the direct component of the stator current is derived, the coefficients of which depend on the torque, velocity, and quadrature current. This equation minimizes the total loss as the sum of copper, iron, and stray losses. The loss minimizing solutions are obtained by a simple numerical approach or using a LUT.

In [20], the MTA control strategy of IPMSM and its flux-weakening control strategy are described. In this chapter, electromagnetic torque and the relationship of the direct and quadrature axis currents can be derived with the curve fitting method directly. The approximation is performed by a second-order polynomial using the method of least squares. In this case, the approximating curve, unlike the approximated one, does not fall into a point with coordinates [0, 0], which reduces the accuracy of regulation in the initial section. Flux-weakening control (FWC) algorithm in this paper is phase shifting. In this case, the system is implemented with two control channels, speed and torque, which complicate the configuration of the system and worsen its dynamic properties.

In [21], a flux-weakening scheme for the IPMSM is proposed. This is done by an additional external voltage regulator of the pulse width modulated (PWM) inverter, which controls the beginning of the flux-weakening and its level.

In [22], a novel field weakening technology of IPMM is described. Here, closed-loop control using the output voltage and feed-forward control with the pre-calculated tables are combined.

In [23], field weakening control of fast dynamics and variable DC-link voltage are achieved by suitable combination of look-up table and voltage feedback controller.

In [24], a voltage-constraint tracking (VCT) field weakening control scheme for IPMSM drives is proposed. The control algorithm is presented in the form of a complicated block diagram with numerous branching and computations. In [25] an approach to minimize the electrical losses of the interior permanent magnet synchronous motors is presented. Two control strategies based on unsaturated and saturated motor model are analyzed. To overcome the problem of parameters unavailable, a procedure is proposed to estimate the parameters of the loss minimization conditions.

In [26], a concept for optimal torque control of IPMSM has been presented. The schema is based on look-up tables, where saturation effects can be considered. A consideration of the permanent magnets' demagnetization effect during flux weakening showed the restrictions between the system parameters.

As the review performed shows, many authors suggest optimizing control systems using pre-calculated tables. The disadvantages of this solution are the reduction of the real-time performance and reliability of the system, increased demands on the amount of processor memory, and the impossibility of adapting the control system to changing parameters and signal disturbances. The use of analytical expressions makes it possible to adapt by on-line identification of the current values of the main parameters. A lot of work is devoted to this problem, for example, [25, 27–31].

Methods based on the use of certain analytical expressions or pre-calculated tables belong to the group "loss model-based control" (LMBC). An alternative for them are the search algorithms, which are most often used in minimizing total losses. Their feature is the measurement of losses or input and output power and the use of iterative methods to find the optimal solution in real time. The search algorithms do not require the knowledge of the motor model and parameters, but in the search process, they have a negative effect on the transients and steady-state values of the controlled coordinates.

This chapter accesses simple speed control field oriented control (FOC) vector systems of IPMSM for constant torque and field weakening operation ranges without using pre-calculated tables, search algorithms, on-line solving of algebraic equations using iterative methods and without the presence of additional control loops. The quality of proposed control systems is investigated via simulation.

The chapter content is structured as follows. In Section 2, the problem is formulated. In Section 3, the IPMSM speed control system in the constant torque operating range is analyzed. In Section 4, speed control system with new field weakening technique is analyzed and designed. Section 5 contains conclusions on the chapter.

2. Problem formulation

For the mathematical description of IPMSM, we use the following symbols: $u_d, u_q, i_d, i_q, \psi_d, \psi_q$— d- and q-axis voltage, current and flux linkage; L_d, L_q—direct and quadrature stator inductance (L_d<L_q); $\Delta L = L_d - L_q$; R—stator resistance; $\tau_d = L_d/R$, $\tau_q = L_q/R$; ω, ω_e—mechanical and electrical rotor speed; p—number of pole pairs; ψ_{pm}—permanent magnet flux linkage; J—moment of inertia; T—motor torque, T_L—load torque.

The differential equations of IPMSM in the d-q rotating reference frame used in the FOC vector systems synthesis are written as follows:

$$\begin{cases} u_d = i_d R + L_d \dfrac{di_d}{dt} - \omega_e \psi_q, \quad \omega_e = p\omega, \quad \psi_q = L_q i_q, \\ u_q = i_q R + L_q \dfrac{di_q}{dt} + \omega_e \psi_d, \quad \psi_d = L_d i_d + \psi_{pm}, \\ T = k_t \left[\psi_{pm} i_q + (L_d - L_q) i_d i_q \right], \quad k_t = 1.5p, \\ J \dfrac{d\omega}{dt} = T - T_L. \end{cases} \tag{1}$$

Table 1 presents the parameters of two motors with different degrees of magnetic asymmetry, for which the research in this chapter is performed.

The task of energy efficient optimal control is to minimize total losses or one of their kinds. Mechanical and switching losses can be most effectively reduced at the design stage of the frequency converter, the motor and the mechanism. Designing a control system based on loss models, usually allows minimizing the copper losses, iron losses or their sum. This chapter discusses methods based on models of copper losses and iron losses from eddy currents.

During analysis and synthesis, the saturation of steel is not taken into account, the methods for parameters identifying and perturbations estimation are not considered.

The graphs in this chapter are presented in p.u. units: $\overline{y} = y/y_b$. As base values, we used: $T_b = T_r$, $i_b = T_r/\left(k_t \psi_{pm}\right)$, $\omega_b = \omega_r = \pi n_r/30$, and $u_b = e_r = p\omega_r \psi_{pm}$.

Parameters	Designation	Values 1	Values 2
Rated speed	n_r, rpm	4000	2000
Rated torque	T_r, Nm	1.8	1.67
Permanent magnet flux linkage	ψ_{pm}, Wb	0.0844	0.0785
Moment of inertia	J, kg m^2	0.45×10^{-3}	0.5×10^{-3}
Number of pole pairs	p	3	2
Stator resistance	R, Ω	2.21	0.87
d-axis inductance	L_d, mH	9.77	14.94
q-axis inductance	L_q, mH	8.72	22.78

Table 1.
Specifications of IPMSM.

3. Speed control of IPMSM in constant torque region

The block diagram of the IPMSM, designed according to Eqs. (1), is shown in **Figure 1**. The synthesis of the vector control system is traditionally performed according to the block diagram shown in **Figure 1**, neglecting the rotation EMF feedback and the crosslinks denoted by dashed lines, as well as the reactive component of the electromagnetic torque denoted by the bold line.

When using PI current controllers (PI-CCs), synthesized by the series correction method, the dotted links are usually compensated by adding corresponding links with opposite signs to the output signals of the PI-CCs. If the asymmetry of the motors magnetic system is not taken into consideration, the proportional gain (P-) and the transfer function of the proportional-integral (PI-) speed controller (SC) are calculated by the formulas:

$$k_{sc} = \frac{i_q^*(s)}{\Delta\omega(s)} = \frac{J}{k_t \psi_{pm} \tau_\omega}, \quad W_{sc}(s) = \frac{i_q^*(s)}{\Delta\omega(s)} = k_{sc} \frac{\tau_{\omega i} s + 1}{\tau_{\omega i} s} \tag{2}$$

where $\tau_\omega = 2\tau_i$ and $\tau_{\omega i} = 2\tau_\omega$; τ_i—integral time-constants of open speed and current loops.

In the first speed range, the copper losses, proportional to the square of the stator current, are determined as:

$$P_{Cu} = 1.5 R i_s^2, \quad i_s^2 = i_q^2 + i_d^2. \tag{3}$$

Then the control problem can be formulated as follows: to find such a relation between the orthogonal components of the stator current, at a given torque, at which the amplitude of the current (2) would be smallest possible. This control strategy is called maximal torque per ampere (MTA).

This optimal control problem is a classical variation task for the conditional extremum, which requires minimizing the expression (3) ensuring the additional torque equation from (1). Such a problem is solved by the Euler-Lagrange method and has a known solution:

$$i_{dMTA} = -\frac{\psi_{pm}}{2(L_d - L_q)} - \sqrt{\frac{\psi_{pm}^2}{4(L_d - L_q)^2} + i_q^2}. \tag{4}$$

In conventional IPMSM FOC system, a reference signal $i_{d0}^* = 0$ is applied to the input of the d-axis stator current (CCd). When implementing the MTA-strategy, the easiest way to take the reference for the q-axis current is from the speed controller and to obtain a reference to the d-axis current from it is through nonlinear functional transformation (4). A fragment of the block diagram for the comparison of the "zero direct current strategy" and the simplest implementation of the MTA strategy is shown in **Figure 2**, where.

$$W_{CCd}(s) = \frac{(\tau_d s + 1)R}{\tau_i s}, \quad W_{CCq}(s) = \frac{(\tau_q s + 1)R}{\tau_i s}.$$

The effectiveness of the MTA strategy is explained in **Figure 3**, which shows in p.u. units a MTA parabola, calculated using expression (4), equal currents circles (Eq. (3)), and equal torques hyperbolae, calculated according to the equation

$$i_q(i_d, T) = \frac{T}{k_t\left[\psi_{pm} + (L_d - L_q)i_d\right]}. \tag{5}$$

The intersection points P_i of these hyperbolae with the MTA trajectory determine the optimal distribution of stator current components for given torque values. Circles with centers at the coordinate system origin, drawn through the P_i points with dashed lines, have radii equal to the total stator currents i_{sMTA} with an optimal distribution of their components.

It can be seen from **Figure 3** that a parabolic MTA trajectory intersects equal torques hyperbolae at an angle of 90°, which provides the minimum possible values

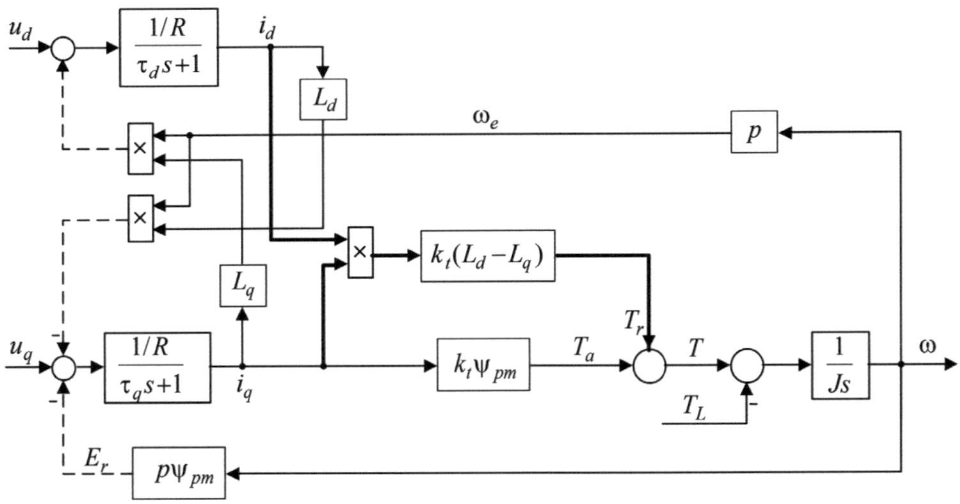

Figure 1.
IPMSM block diagram in rotating rotor reference frame.

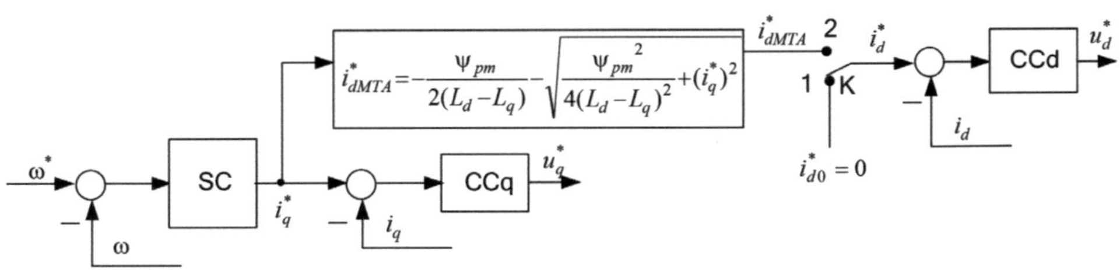

Figure 2.
Block diagram for the comparison of the "zero direct current strategy" and the simplest implementation of MTA strategy.

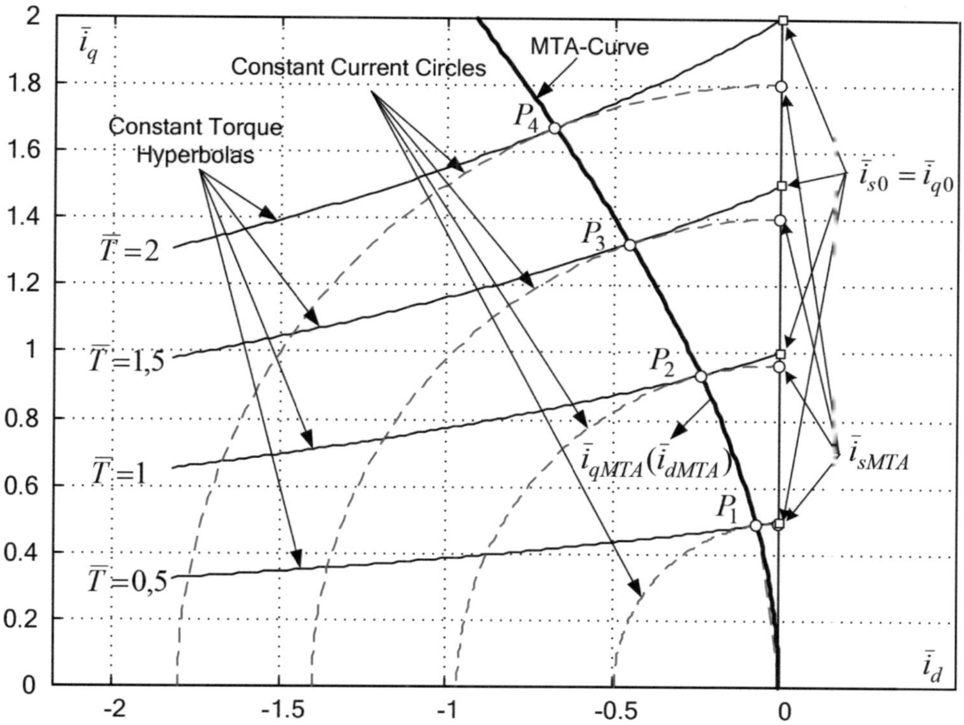

Figure 3.
Current locus diagram.

of the stator current for a given torque. Therefore, the equal torques hyperbolae lie outside the equal currents circles, that is, any ratio of the orthogonal current components that is different from the optimal one leads to an increase in the modulus of the current vector.

To compare currents when using the control strategies under study, attention should be drawn to the cross points of constant currents circles with the q-axis that defines the current modules for the MTA strategy, and the cross points of constant torques hyperbolae with the same axis that defines the peak current for the strategy "$i_d = 0$."

Analysis of the location of these points shows that the advantage of the MTA-strategy increases with increasing electromagnetic torque. This radically differentiates the IPMSM optimal control from the optimal control of induction motor, in which the efficiency of the MTA strategy increases with decreasing electromagnetic torque.

We will perform a study of the static and dynamic properties of IPMSM speed control systems with compared control strategies for the IMPSM with parameters presented in column "Values1" from **Table 1** with time constants $\tau_i = 2\tau_\mu$ = 0.4 ms, $\tau_\omega = 2\tau_i$ = 0.8 ms.

Transient processes during acceleration with the desired electromagnetic torque without load up to the nominal speed and stepwise load torque response for P-SC are shown in **Figure 4**, and for the PI-SC – in **Figure 5**: (a) with "zero d-current strategy"; (b) with the MTA strategy (Eq. (4)).

From the comparison of the transients, it can be seen that in the conventional control system (a) the torque and current curves practically coincide and have the current and torque overshoot values about 5% during the acceleration of and 10% during step disturbance change in the system with P-SC.

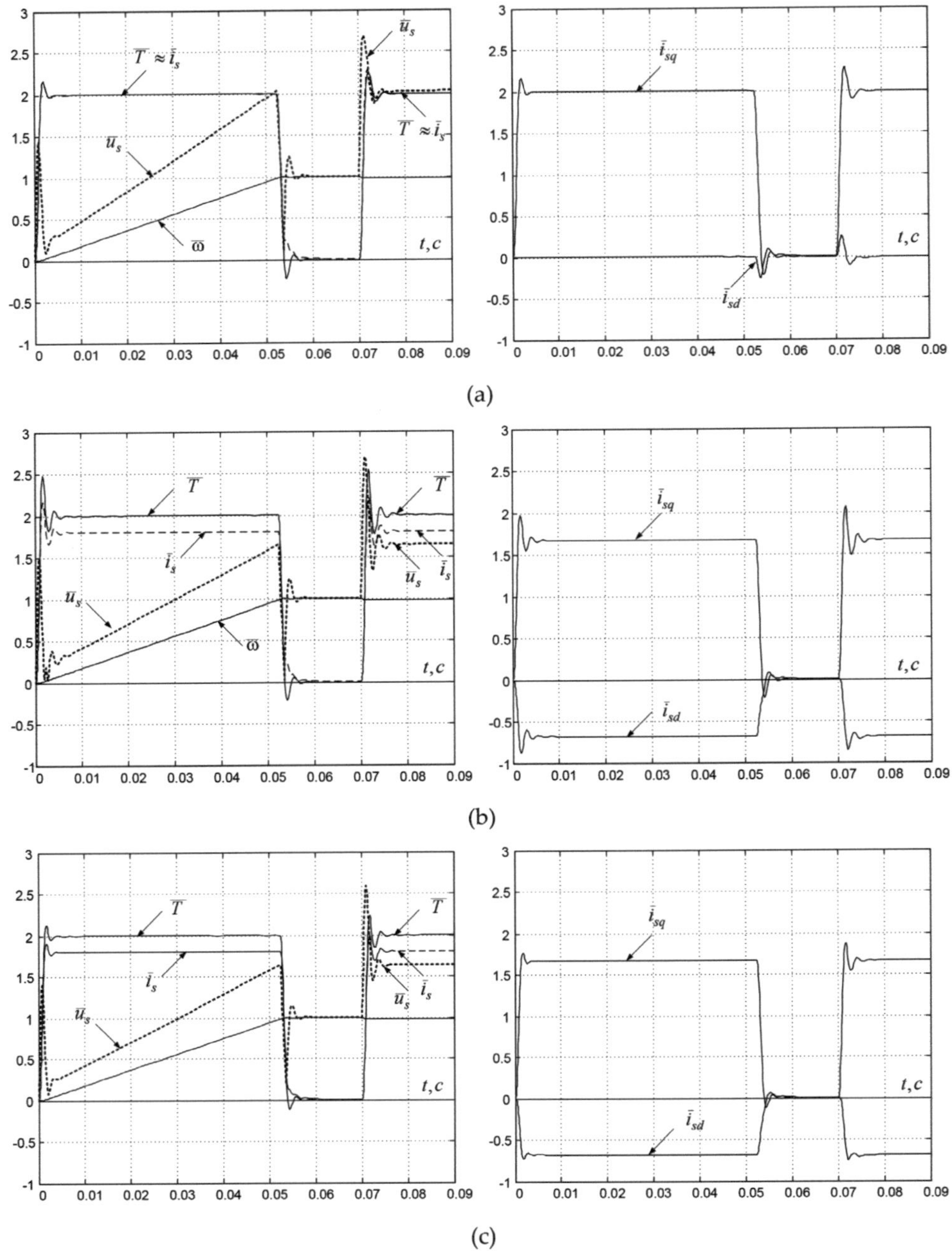

Figure 4.
Transients in IPMSM speed control system with P-SC: (a) used strategy $i_d^* = 0$; (b) used MTA strategy with constant P-SC gain; and (c) used MTA strategy with variable P-SC gain.

In the PI-SC system, the corresponding values are 52 and 65%. The dynamic deviations of the d-current from 0 does not exceed 12 and 25% of the rated value. When applying the MTA strategy without changing the setting of SC adjustments (b), the ratio T/i_s increases in steady-state region, but current and torque overshoots increase up to 25% in during the acceleration and to 35% during step disturbance change in the system from P-SC and to 80 and 90% in a system with PI-SC. In this case, the transient oscillation increases significantly.

From the analysis of the block diagram in **Figure 1**, it follows that this shortcoming is due to incomplete plant compensation by the speed controller. Taking

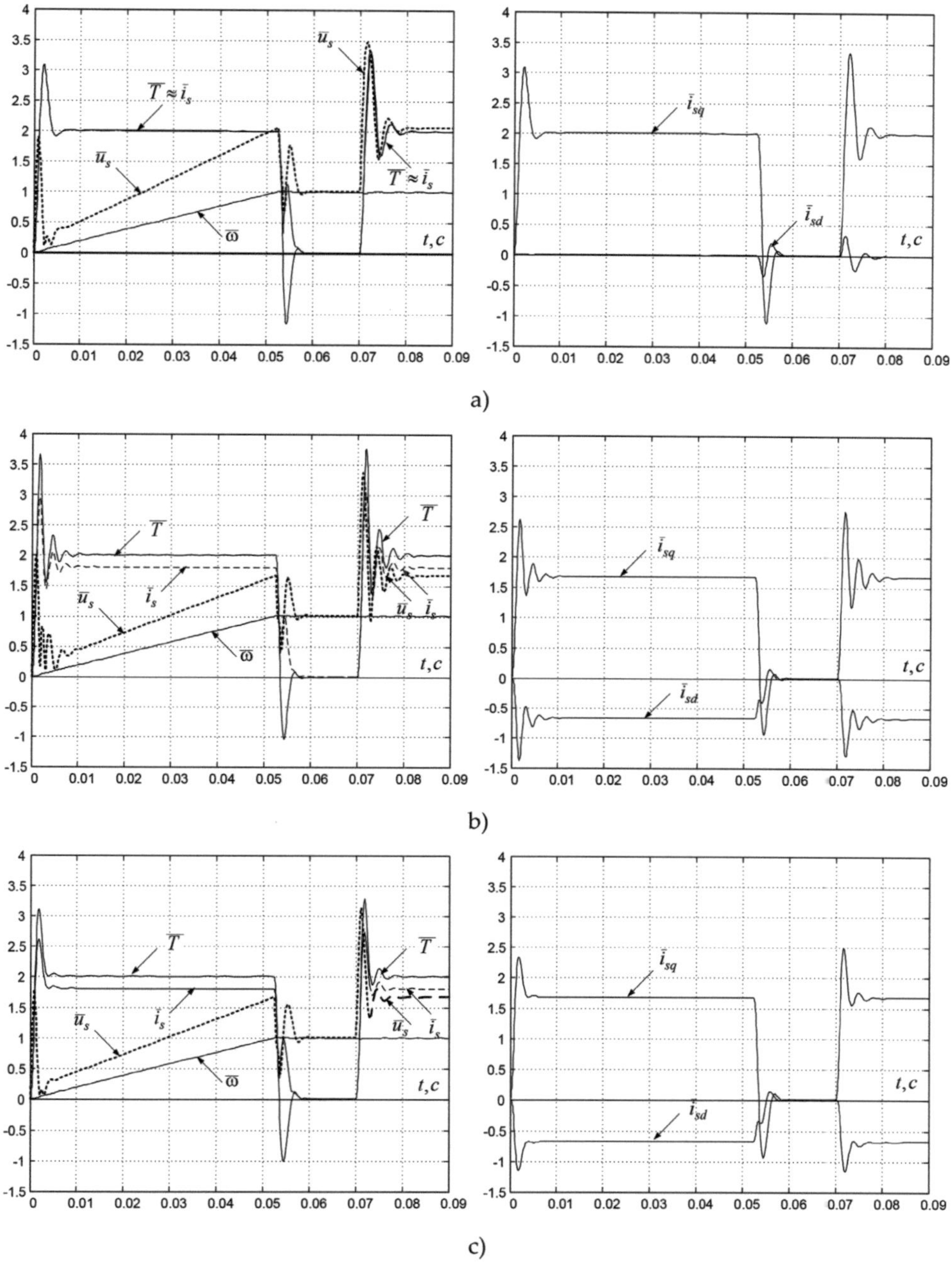

Figure 5.
Transients in IPMSM speed control system with PI-SC: (a) used strategy $i_d^* = 0$; (b) used MTA-strategy without SC adaptation; and (c) used MTA strategy with SC adaptation.

into account of the reactive torque, the P-SC must be designed according to the equation

$$k_{sc1} = \frac{J}{\tau_\omega k_t \left[\psi_{pm} + i_d(t)\left(L_d - L_q\right)\right]} = \frac{J}{\tau_\omega k_t \psi_d(t)}. \tag{6}$$

Transient processes in the investigated system after the proposed speed controller adaptation to reactive torque effect are shown in the **Figure 4c** and **Figure 5c**. They testify the efficiency of the developed technique, since it significantly reduces the torque and currents overshoots, as well as overvoltage and transient oscillations without quality loss of the steady-state energy parameters. Correction of the SC (Eq. 6) improves the dynamics of the system but requires putting the division block after the SC.

Let us consider another approach to the MTA algorithm implementation, in which the SC forms the electromagnetic torque reference. In this case, the P-SC gain and the PI-SC transfer function are calculated by the equations:

$$k_{sc2} = \frac{J}{\tau_\omega}, \quad w_{sc2}(p) = \frac{i_q^*(s)}{\Delta\omega(s)} = k_{sc2}\frac{\tau_{\omega i}s + 1}{\tau_{\omega i}s} \tag{7}$$

If we express d-axis current i_d, through torque T and q-axis current i_q from the 4-th equation of the system (1), and substitute the resulting expression into the left-hand side of Eq. (3), then after transformations, we can write an incomplete fourth degree algebraic equation, that reflects the relationship between the electromagnetic torque and the q-axis stator current in an implicit form:

$$i_{qMTA}{}^4 + \frac{T\psi_{pm}}{k_t(L_d - L_q)^2} i_{qMTA} - \left(\frac{T\psi_{pm}}{k_t(L_d - L_q)}\right)^2 = 0. \tag{8}$$

Eq. (8) can be solved only by numerical methods, which increase the requirements for the microprocessor control devices.

When the system of Eqs. (7), (8) and (4) is used for synthesis, its block diagram acquires the form of **Figure 6**.

To avoid solution of Eq. (8) in real time, the q-axis stator current reference can be calculated using Eq. (5) with $i_q = i_q^*$, $i_d = i_d^*$, and $T = T^*$. In this case, taking into account that, in turn, i_d^* depends on i_q^*, an algebraic loop is formed in a block diagram. This leads to the need to include one sample time delay in the algorithm, as shown in **Figure 7**.

In order to use Eq. (5) in the control algorithm without i_{dMTA}^* signal delay, it is necessary to generate this signal first as a function of the electromagnetic torque. However, the $i_{dMTA}(T)$ dependence is expressed by an even more complicated equation than $i_{qMTA}(T)$:

$$\left(i_{dMTA}{}^2 + \frac{\psi_{pm}}{(L_d - L_q)} i_{do}\right)^2 + \frac{T\psi_{pm}}{k_T(L_d - L_q)^2}\sqrt{i_{dMTA}{}^2 + \frac{\psi_{pm}}{(L_d - L_q)} i_{dMTA}} - \left(\frac{T\psi_{pm}}{k_T(L_d - L_q)}\right)^2 = 0. \tag{9}$$

Therefore, we approximate the curve obtained by the numerical solution of Eq. (9). To do this, we analyze the dependence curves of the orthogonal stator current components and peak values of the stator current from the motor torque for

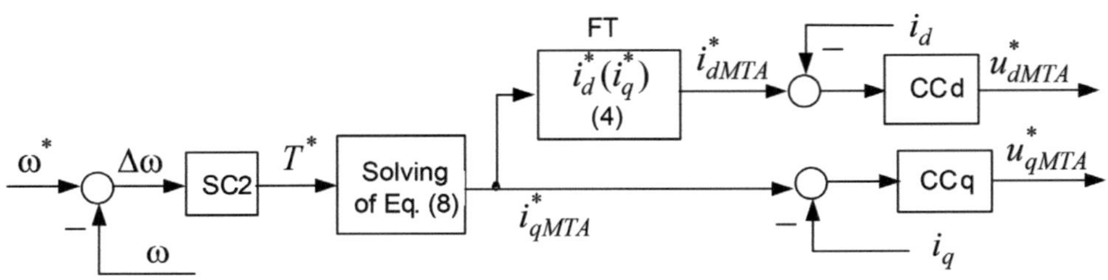

Figure 6.
Fragment of the MTA-optimal block diagram, using Eqs. (8), (4) and SC with a gain (7).

the MTA–and "$i_d = 0$"–strategies for the two IPMSMs from **Table 1**, presented in **Figure 8**. In the latter case, the steady-state value of the peak current coincides with its q-component and is determined from the torque equation:

$$i_{q0}(T) = i_{s0}(T) = \frac{T}{k_m \psi_{pm}}. \tag{10}$$

Figure 8 shows that, firstly, the effectiveness of the MTA strategy increases with the increase of the motor torque and its magnetic asymmetry $|\Delta L| = |L_d - L_q|$ and, secondly, the trajectories $i_{dMTA}(T)$ can be linearized quite easily. For the first motor (**Figure 8a**), the difference between the compared strategies with $T \leq 0.5T_r$ is almost nonexistent, and where this difference becomes significant, the plot $i_{dMTA}(T)$ becomes practically linear. Therefore, the dependence calculated by the Eq. (11) can be replaced by the "dead zone" nonlinearity $i_{dMTAk}(T)$ type, which practically does not change the T/i_s ratio:

$$i_{dMTAk}(T) = \begin{cases} 0 & if \quad |T| \leq T_z, \\ k_l(T - T_z) & if \quad T > T_z, \\ k_l(T + T_z) & if \quad T < -T_z, \end{cases} \tag{11}$$

where T_z—dead band limit and k_l—linear gain.

For the second motor, the nonlinear characteristic can generally be approximated by a straight line drawn through the original curve end points. Finding the parameters of the approximating straight by the least-squares method in this case

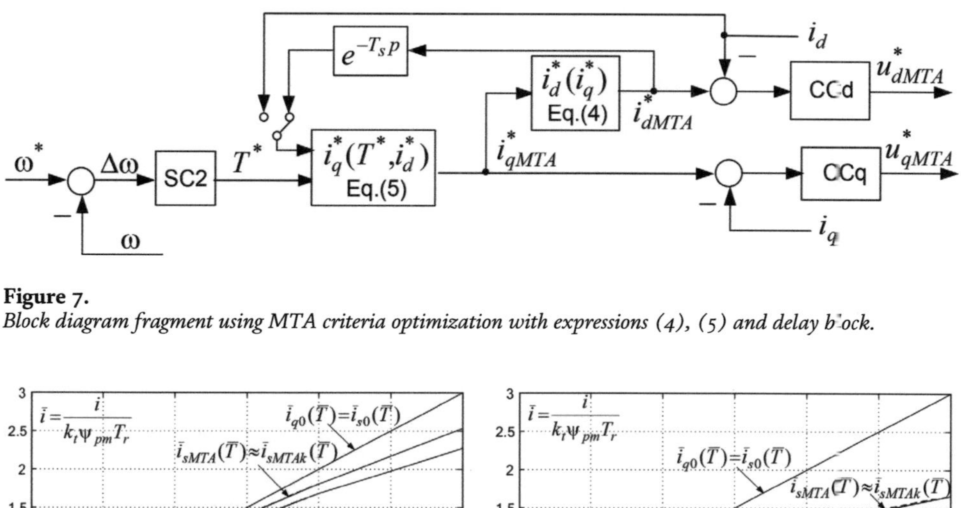

Figure 7.
Block diagram fragment using MTA criteria optimization with expressions (4), (5) and delay block.

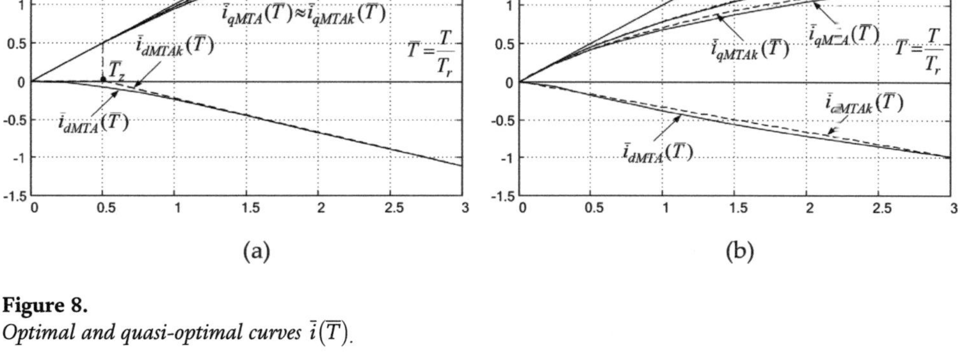

Figure 8.
Optimal and quasi-optimal curves $\bar{i}(\bar{T})$.

makes no sense, since the goal of the approximation is the minimum deviation from the optimal trajectories of not d- or q-components of the stator current, but its amplitude. The approximations obtained for the nonlinear dependence determined by Eq. (11) can be called quasi-optimal. Their use makes it possible to greatly simplify the system with the implementation of the MTA-strategy, as shown in **Figure 9.**

It is known that when using the MTA strategy, the desired torque value is provided not only with a smaller amplitude of the stator current but also with a lower voltage amplitude of the voltage, which further increases the energy efficiency of the control method in question. The plots of the steady-state voltages are shown in **Figure 10**.

To compare the systems under investigation and to confirm the correctness of the plots in **Figures 8** and **10**, we perform simulation using the example of the second motor from **Table 1**.

Transients are shown in **Figure 11** ($i_d = 0$-strategy) and **Figure 12** (quasi-optimal MTA strategy (block diagram **Figure 9**)).

Comparison of the transients shows that in the steady state, the values of the currents, voltages, and their components coincide with the values obtained from the static characteristics in **Figure 8b** and **Figure 10b**. Improvement of energy indicators (reduction of current and voltage amplitudes at the same values of torque and speed) occurs without deteriorating the quality of transient processes. The orthogonal components and voltage amplitude decrease not only for steady-state values of the electromagnetic torque but also during its change.

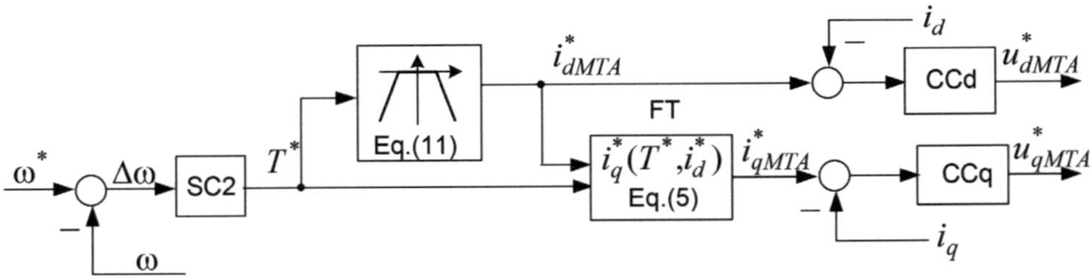

Figure 9.
Block diagram fragment of the MTA-quasi-optimal system.

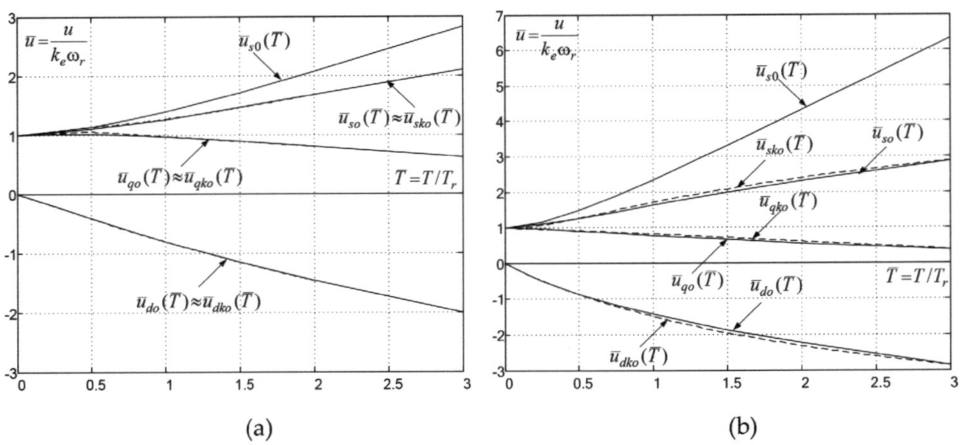

Figure 10.
Optimal and quasi-optimal curves $\bar{u}(\bar{T})$.

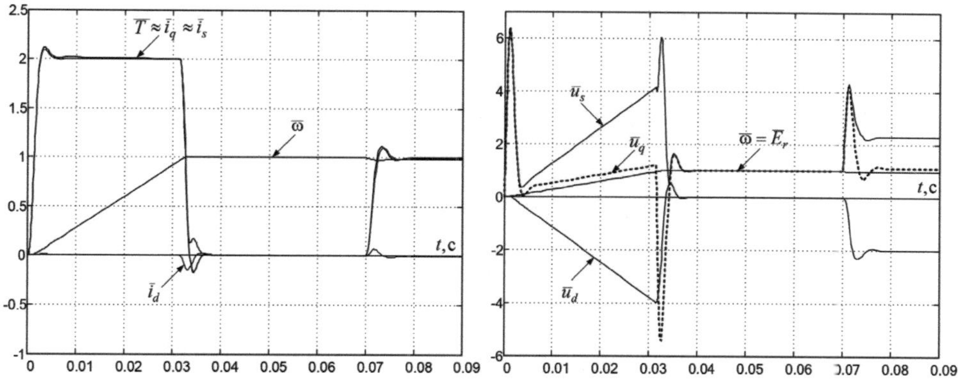

Figure 11.
Transients in the system used $i_d = 0$ strategy.

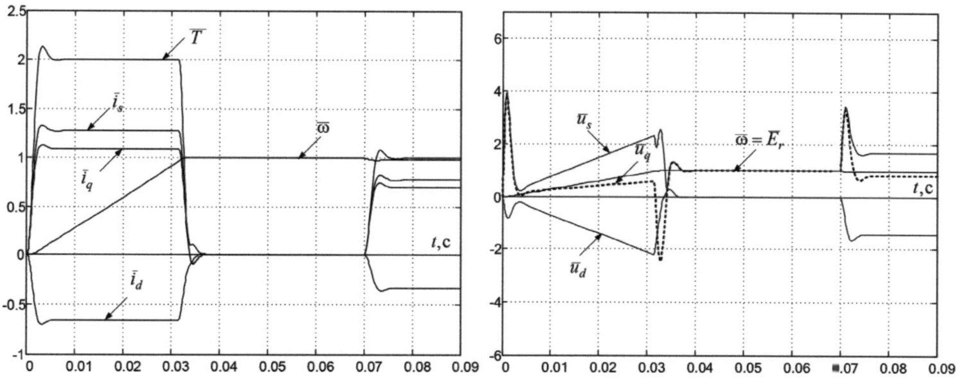

Figure 12.
Transients in the system used quasi-optimal MTA-strategy.

4. Three-range speed control system of IPMSM

The idea of three-range speed regulation follows from the approximated equation for the peak steady-state stator voltage, which can be obtained from the first two equations of system (1), excluding voltage losses on resistance and inductance:

$$u_s^2 = \sqrt{u_q^2 + u_d^2} \approx \omega_e^2 \sqrt{\left(L_q i_q\right)^2 + \left(\psi_{pm} + L_d i_d\right)^2}, \tag{12}$$

from where follows:

$$\omega_e \approx \frac{u_s}{\sqrt{\left(L_q i_q\right)^2 + \left(\psi_{pm} + L_d i_d\right)^2}}. \tag{13}$$

It follows from expression (13) that in the IPMSM, the speed can be increased in three ways:

1. due to a change in the amplitude of the stator voltage from 0 to the nominal value (first range);

2. due to the pseudo-weakening of the permanent magnets field by increasing the d-axis stator current in the negative direction (second range); and

3. due to weakening of the stator field by decreasing the modulus of the q-axis stator current (third range).

The speed control in the first range using the MMA strategy is discussed in the previous section.

The transition to the second range occurs when the stator voltage reaches the nominal value, which is the maximum permissible steady-state voltage value. The relationship between the stator current components in this mode, which is called the field weakening control (FWC), is determined from Eq. (13) with substitutions $u_s = u_{s\max}$ and $\omega_e = \omega_{ep} = \max(\omega_e, p\omega_r)$:

$$i_{dFWC}\left(i_q, u_{s\max}, \omega_{ep}\right) = \frac{-\psi_{pm} + \sqrt{u_{s\max}^2/\omega_{ep}^2 - L_q^2 i_q^2}}{L_d}. \tag{14}$$

Substituting (14) into the torque equation, we obtain implicitly the dependence of the q-axis stator current on the torque, voltage, and speed of the motor [1]:

$$i_{qFWC}^4 + p_2 i_{qFWC}^2 + p_1 i_{qFWC} + p_0 = 0, \tag{15}$$

where

$$p_2 = \frac{\psi_{pm}^2 L_q^2 - \Delta L^2 u_{s\max}^2/\omega_{ep}^2}{L_q^2 \Delta L^2}, \quad p_1 = \frac{4TL_d L_q \psi_{pm}}{3z_p L_q^2 \Delta L^2}, \quad p_0 = \frac{4T^2 L_d^2}{9z_p^2 L_q^2 \Delta L^2}. \tag{16}$$

When adjusting the motor speed in the second range in the current constraint mode, Eq. (16) is modified [1, 2]:

$$i_{dFWC}\left(i_{s\max}, u_{s\max}, \omega_{ep}\right) = \frac{-\psi_{pm} L_q + \sqrt{\psi_{pm}^2 L_q^2 - \left(L_d^2 - L_q^2\right)\left(L_q^2 i_{s\max}^2 + \psi_{pm}^2 - u_{s\max}^2/\omega_{ep}^2\right)}}{L_d^2 - L_q^2}. \tag{17}$$

As the speed increases, iron losses due to eddy currents become more and more significant:

$$P_{Fe} \approx k_{ec} \psi_s^2 \omega_e^2 \approx k_{ec} u_s^2 = 1.5 \cdot u_s^2 / R_{Fe}, \tag{18}$$

where k_{ec} is the eddy current gain, R_{Fe} is the fictitious resistance of the steel, which is inserted into the motor equivalent circuit to simulate this kind of losses.

This makes it advisable to indirectly limit the iron losses by applying the "maximal torque per volt" (MTV), or the "minimal flux per torque" (MFT) control strategies in the third range. Expressing currents through flux linkages

$$i_d = \frac{\psi_d - \psi_{pm}}{L_d}, \quad i_q = \frac{\sqrt{\psi_s^2 - \psi_d^2}}{L_q}$$

and substituting them into the torque equation, we obtain

$$T = \frac{k_t}{L_d L_q} \sqrt{\psi_s^2 - \psi_d^2} \cdot \left[L_q \psi_{pm} - \left(L_d - L_q\right)\psi_d\right].$$

Analyzing the last expression for the extremum, we obtain the following equations that ensure the maximum of the torque:

$$i_{dMTV} = \frac{\psi_{dMTV} - \psi_{pm}}{L_d}, \quad i_{qMTV} = \frac{\sqrt{\left(u_{s\max}/\omega_{ep}\right)^2 - \psi_{dMTV}^2}}{L_q},$$
$$\psi_{dMTV} = \frac{-L_q\psi_{pm} + \sqrt{\psi_{pm}^2 L_q^2 + 8\left(L_d - L_q\right)^2 \left(u_{s\max}/\omega_{ep}\right)^2}}{4\left(L_d - L_q\right)}. \tag{19}$$

Using the Euler–Lagrange equations to find the minimum of the stator voltage (Eq. (12)), taking into account the torque equation as an additional condition, we obtain the dependence between the components of the stator current for the MTV strategy in the following form:

$$i_{dMTV}\left(i_q\right) = -\frac{\psi_{pm}}{2\left(L_d - L_q\right)} \cdot \left(2 - \frac{L_q}{L_d}\right) - \frac{L_q}{L_d}\sqrt{\frac{\psi_{pm}^2}{4\left(L_d - L_q\right)^2} + i_q^2}. \tag{20}$$

In many papers, the current constraint is achieved by limiting the q-axis stator current at the level:

$$\left|i_{q\max i}\right| = i_q\left(i_{s\max}, i_d\right) = \sqrt{i_{s\max}^2 - i_d^2}. \tag{21}$$

Once again, we emphasize that the MTV strategy is applied in the third range, when the reserves for increasing the speed due to the weakening of the field of permanent magnets are exhausted, and regulation is carried out by decreasing the q-component of the stator current, so that the amplitude of the current also decreases.

The p.u. MTA and MTV trajectories calculated using Eqs. (4) and (20) are shown in **Figure 13**. The same graph shows two constant currents circles of

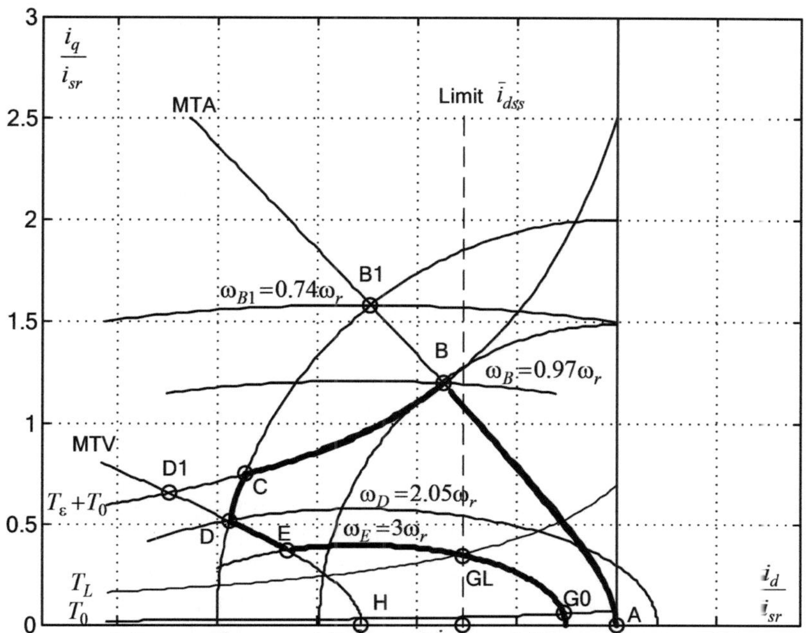

Figure 13.
Steady-state dq-trajectories of IPMSM.

corresponding to Eq. (21), three constant torques hyperbolae, and three constant velocities ellipses, calculated from Eq. (13) with $u_s = u_{s\max}$.

Acceleration begins from point A via almost instantaneous transition to point B or to point B1 on the trajectory of MTA. With a simultaneous increase in the amplitude and frequency of the stator voltage, the change in speed is not accompanied by a transition to another ellipse. After reaching the nominal voltage amplitude value (point B or B1), which determines the upper limit of this value in the steady state $u_s \leq u_{sr} = u_{s\max}$, the execution of MTA strategy becomes impossible.

Transition from the MTA parabola to MTV parabola occurs either along the maximum allowable current circumference (arch B1-D) or a constant torque hyperbola (hyperbolic line B-D1), or first along the constant torque hyperbola with increase of the current, and then along the overcurrent circumference (path BCD). Movement along the hyperbola is carried out with an underloaded motor with constant torque and voltage. The power increases due to the increase in speed and current. Movement along the circumference occurs at a constant power. These trajectories control the speed in the second range in accordance with the FWC strategy (Eq. 14–17).

Further increase in the speed to the desired value occurs in the third range along the MTV trajectory (Eq. (19)). In this mode (segment D-E), the orthogonal components of the stator current are reduced, resulting in the motor coming out of the current-limiting mode and operating at the reduced values of both stator current amplitude and power.

After the speed has reached its set value, the transition from the acceleration mode to the steady state occurs. In this case, the motion of the working point occurs along an ellipse of constant velocity until it intersects with the load torque hyperbola (segment E-GL or E-G0).

Concentric ellipses of constant velocities have a center at the point H with coordinates $i_{d0} = \psi_{pm}/L_d$, $i_{q0} = 0$, in which the MTV line ends. At this point, a complete demagnetization of the motor takes place, which theoretically makes it possible to achieve an arbitrarily large steady-state velocity ($\omega \to \infty$) at zero-load torque ($T_L = 0$). In practice, these points are unattainable, and the real range of speed control in the third range is limited by the magnitude of the load torque, the mechanical strength of the rotor, and the constraint of the d-axis stator current in the steady state at about $i_{d0}/2$ to prevent the permanent magnets from being irreversibly demagnetized.

As can be seen from the above Eqs. (4, 5, 8, 9, 14–17, 19–21), a general algorithm for controlling the speed of IPMSM in the three range requires significant computational resources associated with the need for a numerical solution of algebraic equations in real time and complex logic branching for the purpose of organizing controlled switching, etc. Replacing a single algorithm with multidimensional look-up tables of data is associated with the preliminary calculation of a large number of curves and the organization of the search in these tables.

Meanwhile, the analysis of the control trajectories in **Figure 13** suggests the possibility of a structural implementation of the control algorithm, which is presented in **Figure 14**.

In it, the speed controller SC forms the torque reference. To avoid solving fourth-degree algebraic equations in real time (Eqs. 8, 9, 15) and computations according to formulas (19), the q-axis stator current reference is determined by the Eq. (5) with $T = T^*$, $i_q = i_q{}^*$. To prevent the formation of an algebraic loop, we use in this equation not the reference $i_d{}^*$, but the feedback signal i_d.

The limitation of the stator current amplitude is achieved by the dynamic saturation block Sat1 for the q-axis current at the level (21), which is calculated

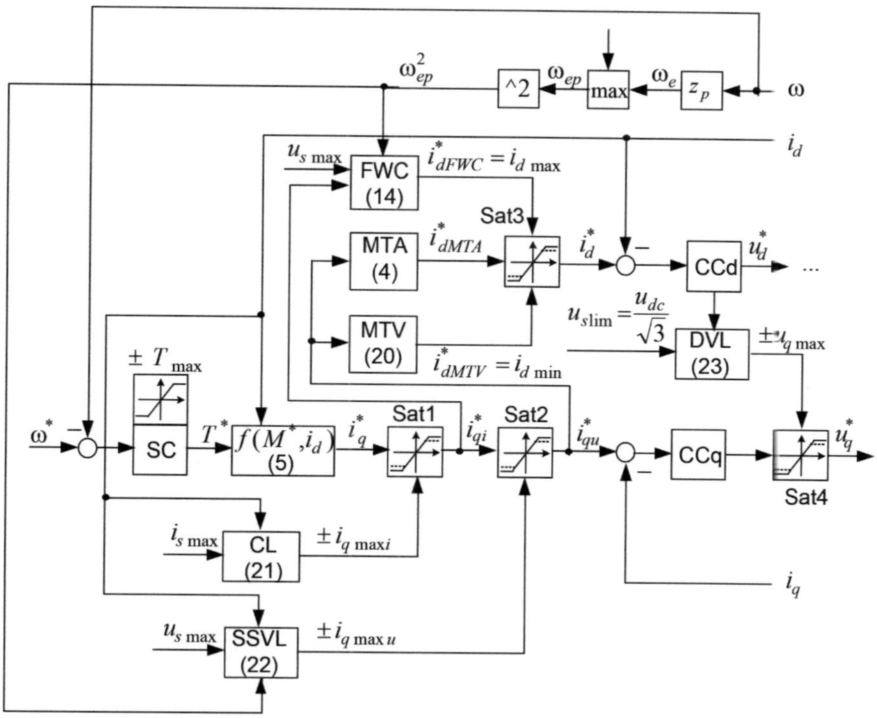

Figure 14.
Block diagram of the three-range IPMSM speed control system considering voltage and current constrains.

by current limit block CL. The dynamic saturation block Sat2 for the stator voltage in no-load mode is switched in series with the block Sat1, and the constraint signal is formed by the steady-state voltage limit block SSVL according to the equation

$$i_{q\max u} = \sqrt{\left(u_{s\max}/\omega_{ep}\right)^2 - \left(L_d i_d + \psi_{pm}\right)^2}/L_q, \tag{22}$$

obtained from Eq. (12) at $u_s = u_{s\max}$. Thus, at the output of the block Sat2, we obtain a signal $\left|i_{qu}\right| = \min\left[\left|i_q\right|, \left|i_{q\max i}\right|, \left|i_{q\max u}\right|\right]$.

To ensure that the motor voltage in dynamic modes does not exceed the inverter DC link voltage, the output of the regulator CCq is limited at the level

$$u_s \le u_{\lim} = u_{dc}/\sqrt{3} \tag{23}$$

with the blocks of dynamic voltage limit (DVL) and Sat4.

The d-axis current reference signal in the first region is calculated from the Eq. (4) as a function of the signal (22), in the second range from the Eq. (14) as the function of signal (21) and in the third region from the Eq. (20) with the input signal (22). The switching of the control algorithm from MTA to FWC and from FWC to MTV occurs by dynamically limiting the signal of the MTA block at the top level $i_{dFWC}{}^*$ and at the bottom level $i_{dMTV}{}^*$.

Transients in the system shown in **Figure 14**, obtained via simulation, are shown in **Figure 15**.

The coordinates of the characteristic points of the transients in **Figure 15** coincide with the corresponding points of the diagrams in **Figure 13**. The stator current does not exceed its maximum permissible value, and the stator voltage in no-load

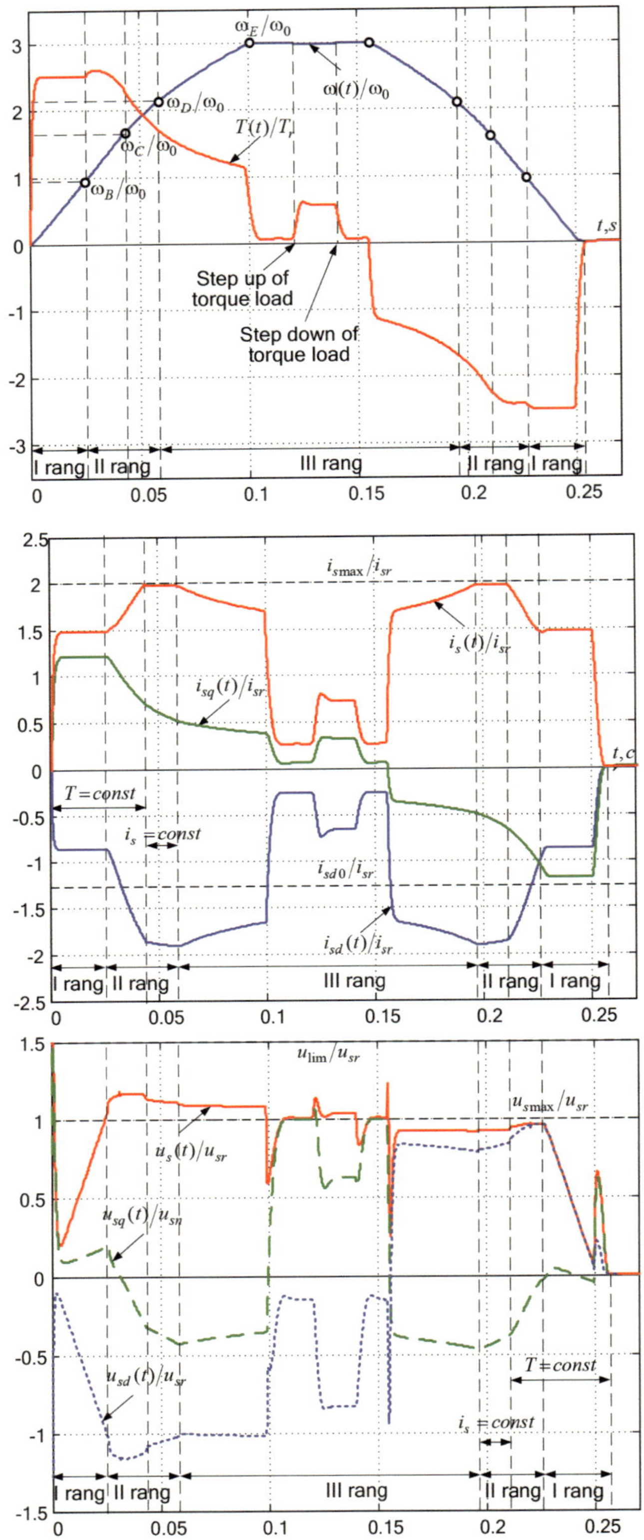

Figure 15.
Transients in the 3-range speed control system of IPMSM.

mode is equal to the nominal stator voltage, and in dynamic modes, it is limited by the output voltage of the inverter DC link.

5. Conclusions

In this chapter, we presented the IPMSM speed control systems. Two options for improving the quality of the transient processes for the speed regulation by changing the stator voltage using the MTA strategy are proposed. In a system in which the speed controller generates q-axis stator current reference, a method is proposed for adapting the speed controller to the presence of the reactive component of the electromagnetic torque. In a system in which the speed controller forms the electromagnetic torque reference, it is suggested to approximate the dependence $i_d^*(T^*)$ using the "dead zone" nonlinearity and thus avoid solving the fourth-order equation in real time. At the same time, it was possible to ensure high energy parameters without the use of pre-calculated tables and organization of search in them without degrading the quality of transient processes.

For the system with a three-range speed regulation, a system is proposed using MTA, FWC, and MTV strategies with automatic switching between them and stator current and voltage constraints without using additional control loops is proposed.

The analytical researches are confirmed by the simulation results.

Author details

Olga Tolochko
National Technical University of Ukraine, Igor Sikorsky Polytechnic Institute, Kyiv, Ukraine

*Address all correspondence to: tolochko.ola@gmail.com

References

[1] Schröder D. Elektrische Antriebe – Regelung von Antriebssystemen. 3. bearbeitete Auflage. Berlin, Heidelberg: Springer; 2009. p. 1336. DOI: 10.1007/978-3-540-89613-5

[2] Krishnan R. Permanent magnet synchronous and brushless DC motor drives. Virginia Tech, Blacksburg, USA: CRC press; 2010. p. 611. ISBN 978-0-8247-5384-9

[3] Bose BK. Modern power electronics and AC drives. New Jersey: Prentice Hall PTR; 2002. p. 711. ISBN 0130167436, 9780130167439

[4] Sul S-K. Control of electric machine drive systems. Willey-IEEE Press; 2011. p. 424. ISBN: 978-0-470-59079-9

[5] Doncker RD, Pulle DWJ, Veltman A. Advanced electrical drives. Analysis, modeling, control. Berlin: Springer; 2011. p. 455. DOI: 10.1007/978-94-007-0181-6

[6] Agrawal J, Bodkhe S. Steady-state analysis and comparison of control strategies for PMSM. Modelling and Simulation in Engineering; 2015. p. 11. DOI: 10.1155/2015/306787

[7] Morimoto S, Takeda Y, Hatanaka K, Tong Y, Hirasa T. Design and control system of permanent magnet synchronous motor for high torque and high efficiency operation. In: Industry Applications Conference Record of the 1991 IEEE Industry; 1991; 1. pp. 176-181. DOI: 10.1109/IAS.1991.178151

[8] Morimoto S, Hatanaka K, Tong Y, Takeda Y, Hirasa T. High performance servo drive system of salient pole permanent magnet synchronous motor. In: Industry Applications Society Annual Meeting of the IEEE Conference Record Proceedings; 1; 28 September–4 October 1991; pp. 463-468. DOI: 10.1109/IAS.1991.178196

[9] Morimoto S, Sanada M, Takeda Y. Wide speed operation of interior permanent magnet synchronous motors with high performance current regulator. IEEE Transactions on Industry Applications. 1994;**IA-30**(4): 920-926. DOI: 10.1109/28.297908

[10] Morimoto S, Hatanaka K, Tong Y, Takeda Y, Hirasa T. Servo drive system and control characteristics of salient pole permanent magnet synchronous motor. IEEE Transactions on Industry Applications. 1993;**29**(2):338-343. DOI: 10.1109/28.216541

[11] Lee J-G, Nam K-H, Lee S-H, Choi S-H, Kwon S-W. A lookup table based loss minimizing control for FCEV permanent magnet synchronous motors. In: 2007 IEEE Vehicle Power and Propulsion Conference Journal of Electrical Engineering Technology; 9–12 September 2007; Arlington, USA; 4(2). pp. 201-210. DOI: 10.1109/VPPC.2007.4544120

[12] Huang S, Chen Z, Huang K, Gao J. Maximum torque per ampere and flux-weakening control for PMSM based on curve fitting. In: Vehicle Power and Propulsion Conference (VPPC), 2010 IEEE; 1–3 September 2010; Lille, France. DOI: 10.1109/VPPC.2010.5729024

[13] Ma L, Sanada M, Morimoto S, Takeda Y. Prediction of iron loss in rotating machines with rotational loss included magnetics. In: IEEE Transactions on Magnetics; 2003. 39(4). pp. 2036-2041. DOI: 10.1109/TMAG.2003.812706

[14] Li L, Huang X, Baoquan Kao B, Yan B. Research of core loss of permanent

magnet synchronous motor (PMSM) in AC servo system. In: Electrical Machines and Systems. ICEMS; 17–20 October 2008; Wuhan, China. pp. 602-607. INSPEC Accession Number: 10458526

[15] Navrapescu V, Kisck DO, Popescu M, Kisck M, Andronescu G. Modeling of iron losses in salient pole permanent magnet synchronous motors. In: 7th International Conference on Power Electronics (ICPE '07); 22–26 October 2007; Daegu, South Korea. pp. 352-357. DOI: 10.1109/ICPE.2007.4692408

[16] Urasaki N, Senjyu T, Uezato K. A novel calculation method for iron loss resistance suitable in modeling permanent-magnet synchronous motors. IEEE Transactions on Energy Conversion. 2003;**18**(1):41-47. DOI: 10.1109/TEC.2002.808329

[17] Wijenayake AH, Schmidt PB. Modeling and analysis of permanent magnet synchronous motor by taking saturation and core loss into account. In: 1997 International Conference on Power Electronics and Drive Systems; 26–29 May 1997; Singapore. DOI: 10.1109/PEDS.1997.627416

[18] Yan Y, Zhu J, Guo Y. A permanent magnet synchronous motor model with core loss. Journal of the Japan Society of Applied Electromagnetic and Mechanics. 2007;**15**:147-150

[19] Mansouri A, Trabelsi H. On the performances investigation and iron losses computation of an inset surface mounted permanent magnet motor. Systems, Signals and Devices. 2012. P. 1-5

[20] Mi C, Slemo GR, Bonert R. Minimization of iron losses of permanent magnet synchronous machines. IEEE Transactions on Energy Conversion. 2005;**20**:121-127. DOI: 10.1109/TEC.2004.832091

[21] Kim J.-M., Sul S.-K. Speed control of interior permanent magnet synchronous motor drive for the flux weakening operation. In: IEEE Transactions on Industry Applications, Vol. 33, No. 1, January/February 1997. p. 43–48. DOI: 10.1109/28.567075

[22] Bae B-H, Patel N, Schulz S, Sul S-K. New field weakening technique for high saliency interior permanent magnet motor. In: 38th IAS Annual Meeting on Industry Applications Conference, 2003; 12–16 October 2003; Salt Lake City, USA. pp. 898-905. DOI: 10.1109/IAS.2003.1257641

[23] Lee JH, Lee JH, Park JH, Won CY. Field-weakening strategy in condition of DC-link voltage variation using on electric vehicle of IPMSM. In: Electrical Machines and Systems (ICEMS). 20–23 August 2011; Beijing, China; 2011. 27 Jan/Feb. pp. 38-44. DOI: 10.1109/ICEMS.2011.6073676

[24] Sue S-M, Pan C-T. Voltage-constraint-tracking-based field-weakening control of IPM synchronous motor drives. In: IEEE Transactions on Industrial Electronics; 55(1); January 2008. pp. 340-347. DOI: 10.1109/TIE.2007.909087

[25] Hao S, Shi J, Hao M, Mizugaki Y. Closed-loop parameter identification of permanent magnet synchronous motor considering nonlinear influence factors. Journal of Advanced Mechanical Design, Systems and Manufacturing. 2010;**4**(6):1157-1165. DOI: 10.1299/jamdsm.4.1157

[26] Meyer M, Bocker J. Optimum control for interior permanent magnet synchronous motors (IPMSM) in constant torque and flux weakening range. In: Power Electronics and Motion Control Conference, 2006. EPE-PEMC 2006. 12th International Power Electronics and Motion Control

Conference; 10 February 2006; Portoroz, Slovenia. pp. 282-286. DOI: 10.1109/EPEPEMC.2006.4778413

[27] Inoue Y, Yamada K, Morimoto S, Sanada M. Effectiveness of voltage error compensation and parameter identification for model-based sensorless control of IPMSM. IEEE Transactions on Industry Applications. 2009;**45**(1):213-221. DOI: 10.1109/TIA.2008.2009617

[28] Mink F, Kubasiak N, Ritter B, Binder A. Parametric model and identification of PMSM considering the influence of magnetic saturation. In: 13th International Conference on Optimization of Electrical and Electronic Equipment. 24–26 May 2012; Brasov, Romania; 2012. pp. 444-452. DOI: 10.1109/OPTIM.2012.6231768

[29] Senjyu T, Kinjo K, Urasaki N, Uezato K. Parameter measurement for PMSM using adaptive identification. In: Industrial Electronics, 2002. ISIE 2002. Proceedings of the 2002 IEEE International Symposium; 3. pp. 711-716

[30] Tolochko OI, Bozhko VV. Inductances identification of IPMSM by the recurrent method of least squares. Scientific papers of Donetsk National Technical University. Series: Electrical engineering and power engineering; 2012;**12**(2):234-238

[31] Trandafilov VN, Bozhko VV, Tolochko OI. Inertia identification of electric drive by an unnormalized gradient method. Scientific papers of Donetsk National Technical University. Series: Electrical engineering and power engineering. 2012;**10**(180):194-199

Chapter 9

Convexity, Majorization and Time Optimal Control of Coupled Spin Dynamics

Navin Khaneja

Abstract

In this chapter, we study some control problems that derive from time optimal control of coupled spin dynamics in NMR spectroscopy and quantum information and computation. Time optimal control helps to minimize relaxation losses. In a two qubit system, the ability to synthesize, local unitaries, much more rapidly than evolution of couplings, gives a natural time scale separation in these problems. The generators of unitary evolution, $\mathfrak{g}$, are decomposed into fast generators $\mathfrak{k}$ (local Hamiltonians) and slow generators $\mathfrak{p}$ (couplings) as a Cartan decomposition $\mathfrak{g} = \mathfrak{p} \oplus \mathfrak{k}$. Using this decomposition, we exploit some convexity ideas to completely characterize the reachable set and time optimal control for these problems. The main contribution of the chapter is, we carry out a global analysis of time optimality.

Keywords: Kostant convexity, spin dynamics, Cartan decomposition, Cartan subalgebra, Weyl group, time optimal control

1. Introduction

A rich class of model control problems arise when one considers dynamics of two coupled spin $\frac{1}{2}$. The dynamics of two coupled spins, forms the basis for the field of quantum information processing and computing [1] and is fundamental in multidimensional NMR spectroscopy [2, 3]. Numerous experiments in NMR spectroscopy, involve synthesizing unitary transformations [4–6] that require interaction between the spins (evolution of the coupling Hamiltonian). These experiments involve transferring, coherence and polarization from one spin to another and involve evolution of interaction Hamiltonians [2]. Similarly, many protocols in quantum communication and information processing involve synthesizing entangled states starting from the separable states [1, 7, 8]. This again requires evolution of interaction Hamiltonians between the qubits.

A typical feature of many of these problems is that evolution of interaction Hamiltonians takes significantly longer than the time required to generate local unitary transformations (unitary transformations that effect individual spins only). In NMR spectroscopy [2, 3], local unitary transformations on spins are obtained by application of rf-pulses, whose strength may be orders of magnitude larger than the couplings between the spins. Given the Schróedinger equation for unitary evolution

$$\dot{U} = -i\left[H_c + \sum_{j=1}^{n} u_j H_j\right]U, \quad U(0) = I, \tag{1}$$

where H_c represents a coupling Hamiltonian, and u_j are controls that can be switched on and off. What is the minimum time required to synthesize any unitary transformation in the coupled spin system, when the control generators H_j are local Hamiltonians and are much stronger than the coupling between the spins (u_j can be made large). Design of time optimal rf-pulse sequences is an important research subject in NMR spectroscopy and quantum information processing [4, 9–21], as minimizing the time to execute quantum operations can reduce relaxation losses, which are always present in an open quantum system [22, 23]. This problem has a special mathematical structure that helps to characterize all the time optimal trajectories [4]. The special mathematical structure manifested in the coupled two spin system, motivates a broader study of control systems with the same properties.

The Hamiltonian of a spin $\frac{1}{2}$ can be written in terms of the generators of rotations on a two dimensional space and these are the Pauli matrices $-i\sigma_x, \ -i\sigma_y, \ -i\sigma_z$, where,

$$\sigma_z = \frac{1}{2}\begin{bmatrix} 1 & 0 \\ 0 & -1 \end{bmatrix}; \ \sigma_y = \frac{1}{2}\begin{bmatrix} 0 & -i \\ i & 0 \end{bmatrix}; \ \sigma_x = \frac{1}{2}\begin{bmatrix} 0 & 1 \\ 1 & 0 \end{bmatrix}. \tag{2}$$

Note

$$[\sigma_x, \sigma_y] = i\sigma_z, \quad [\sigma_y, \sigma_z] = i\sigma_x, \quad [\sigma_z, \sigma_x] = i\sigma_y, \tag{3}$$

where $[A, B] = AB - BA$ is the matrix commutator and

$$\sigma_x^2 = \sigma_y^2 = \sigma_z^2 = \frac{\mathbf{1}}{4}, \tag{4}$$

The Hamiltonian for a system of two coupled spins takes the general form

$$H_0 = \sum a_\alpha \sigma_\alpha \otimes \mathbf{1} \ + \sum b_\beta \mathbf{1} \ \otimes \sigma_\beta + \sum J_{\alpha\beta} \ \sigma_\alpha \otimes \sigma_\beta, \tag{5}$$

where $\alpha, \beta \in \{x, y, z\}$. The Hamiltonians $\sigma_\alpha \otimes \mathbf{1}$ and $\mathbf{1} \ \otimes \sigma_\beta$ are termed local Hamiltonians and operate on one of the spins. The Hamiltonian

$$H_c = \sum J_{\alpha\beta} \ \sigma_\alpha \otimes \sigma_\beta, \tag{6}$$

is the coupling or interaction Hamiltonian and operates on both the spins. The following notation is therefore common place in the NMR literature.

$$I_\alpha = \sigma_\alpha \otimes \mathbf{1}; \quad S_\beta = \mathbf{1} \ \otimes \sigma_\beta. \tag{7}$$

The operators I_α and S_β commute and therefore $\exp\left(-i\sum_\alpha a_\alpha I_\alpha + \sum_\beta b_\beta S_\beta\right) =$

$$\exp\left(-i\sum_\alpha a_\alpha I_\alpha\right)\exp\left(-i\sum_\beta b_\beta S_\beta\right) = \left(\exp\left(-i\sum_\alpha a_\alpha \sigma_\alpha\right)\otimes \mathbf{1}\right)\left(\mathbf{1} \ \otimes \exp\left(-i\sum_\beta b_\beta \sigma_\beta\right)\right), \tag{8}$$

The unitary transformations of the kind

$$\exp\left(-i\sum_\alpha a_\alpha \sigma_\alpha\right)\otimes \exp\left(-i\sum_\beta b_\beta \sigma_\beta\right),$$

obtained by evolution of the local Hamiltonians are called local unitary transformations.

The coupling Hamiltonian can be written as

$$H_c = \sum J_{\alpha\beta} I_\alpha S_\beta. \tag{9}$$

Written explicitly, some of these matrices take the form

$$I_z = \sigma_z \otimes \mathbf{1} = \frac{1}{2}\begin{bmatrix} 1 & 0 & 0 & 0 \\ 0 & 1 & 0 & 0 \\ 0 & 0 & -1 & 0 \\ 0 & 0 & 0 & -1 \end{bmatrix}. \tag{10}$$

and

$$I_z S_z = \sigma_z \otimes \sigma_z = \frac{1}{4}\begin{bmatrix} 1 & 0 & 0 & 0 \\ 0 & -1 & 0 & 0 \\ 0 & 0 & -1 & 0 \\ 0 & 0 & 0 & 1 \end{bmatrix}. \tag{11}$$

The 15 operators,

$$-i\{I_\alpha, S_\beta, I_\alpha S_\beta\},$$

for $\alpha, \beta \in \{x, y, z\}$, form the basis for the Lie algebra $\mathfrak{g} = su(4)$, the 4×4, traceless skew-Hermitian matrices. For the coupled two spins, the generators $-iH_c, -iH_j \in su(4)$ and the evolution operator $U(t)$ in Eq. (1) is an element of $SU(4)$, the 4×4, unitary matrices of determinant 1.

The Lie algebra $\mathfrak{g} = su(4)$ has a direct sum decomposition $\mathfrak{g} = \mathfrak{p} \oplus \mathfrak{k}$, where

$$\mathfrak{k} = -i\{I_\alpha, S_\beta\}, \quad \mathfrak{p} = -i\{I_\alpha S_\beta\}. \tag{12}$$

Here $\mathfrak{k}$ is a subalgebra of $\mathfrak{g}$ made from local Hamiltonians and $\mathfrak{p}$ nonlocal Hamiltonians. In Eq. (1), we have $-iH_j \in \mathfrak{k}$ and $-iH_c \in \mathfrak{p}$, It is easy to verify that

$$[\mathfrak{k}, \mathfrak{k}] \subset \mathfrak{k}, \quad [\mathfrak{k}, \mathfrak{p}] \subset \mathfrak{p}, \quad [\mathfrak{p}, \mathfrak{p}] \subset \mathfrak{p}. \tag{13}$$

This decomposition of a real semi-simple Lie algebra $\mathfrak{g} = \mathfrak{p} \oplus \mathfrak{k}$ satisfying (13) is called the Cartan decomposition of the Lie algebra $\mathfrak{g}$ [24].

This special structure of Cartan decomposition arising in dynamics of two coupled spins in Eq. (1), motivates study of a broader class of time optimal control problems.

Consider the following canonical problems. Given the evolution

$$\dot{U} = \left(X_d + \sum_j u_j(t) X_j\right) U, \quad U(0) = \mathbf{1}\ , \tag{14}$$

where $U \in SU(n)$, the special Unitary group (determinant 1, $n \times n$ matrices U such that $UU' = \mathbf{1}$, $'$ is conjugate transpose). Where $X_j \in \mathfrak{k} = so(n)$, skew symmetric matrices and

$$X_d = -i\begin{bmatrix} \lambda_1 & 0 & \dots & 0 \\ 0 & \lambda_2 & \dots & 0 \\ \vdots & \vdots & \ddots & \vdots \\ 0 & 0 & \dots & \lambda_n \end{bmatrix}, \quad \sum \lambda_i = 0.$$

We assume $\{X_j\}_{LA}$, the Lie algebra (X_j and its matrix commutators) generated by generators X_j is all of $so(n)$. We want to find the minimum time to steer this system between points of interest, assuming no bounds on our controls $u_j(t)$. Here again we have a Cartan decomposition on generators. Given $\mathfrak{g} = su(n)$, traceless skew-Hermitian matrices, generators of $SU(n)$, we have $\mathfrak{g} = \mathfrak{p} \oplus \mathfrak{k}$, where $\mathfrak{p} = -iA$, where A is traceless symmetric and $\mathfrak{k} = so(n)$. As before, $X_d \in \mathfrak{p}$ and $X_j \in \mathfrak{k}$. We want to find time optimal ways to steer this system. We call this $\frac{SU(n)}{SO(n)}$ problem. For $n = 4$, this system models the dynamics of two coupled nuclear spins in NMR spectroscopy.

In general, U is in a compact Lie group G (such as $SU(n)$), with X_d, X_j in its real semisimple (no abelian ideals) Lie algebra $\mathfrak{g}$ and

$$\dot{U} = \left(X_d + \sum_j u_j(t) X_j \right) U, \quad U(0) = \mathbf{1} \ . \tag{15}$$

Given the Cartan decomposition $\mathfrak{g} = \mathfrak{p} \oplus \mathfrak{k}$, where $X_d \in \mathfrak{p}$, $\{X_j\}_{LA} = \mathfrak{k}$ and $K = \exp(\mathfrak{k})$ (product of exponentials of $\mathfrak{k}$) a closed subgroup of G, We want to find the minimum time to steer this system between points of interest, assuming no bounds on our controls $u_j(t)$. Since $\{X_j\}_{LA} = \mathfrak{k}$, any rotation (evolution) in subgroup K can be synthesized with evolution of X_j [25, 26]. Since there are no bounds on $u_j(t)$, this can be done in arbitrarily small time [4]. We call this $\frac{G}{K}$ problem.

The special structure of this problem helps in complete description of the reachable set [27]. The elements of the reachable set at time T, takes the form $U(T) \in$

$$S = K_1 \exp\left(T \sum_k \alpha_k \ \mathcal{W}_k X_d \mathcal{W}_k^{-1} \right) K_2, \tag{16}$$

where $K_1, K_2, \mathcal{W}_k \in \exp(\mathfrak{k})$, and $\mathcal{W}_k X_d \mathcal{W}_k^{-1}$ all commute, and $\alpha_k > 0$, $\sum \alpha_k = 1$. This reachable set is formed from evolution of K_1, K_2 and commuting Hamiltonians $\mathcal{W}_k X_d \mathcal{W}_k^{-1}$. Unbounded control suggests that $K_1, K_2, \mathcal{W}_k$ can be synthesized in negligible time.

This reachable set can be understood as follows. The Cartan decomposition of the Lie algebra $\mathfrak{g}$, in Eq. (13) leads to a decomposition of the Lie group G [24]. Inside $\mathfrak{p}$ is contained the largest abelian subalgebra, denoted as $\mathfrak{a}$. Any $X \in \mathfrak{p}$ is Ad_K conjugate to an element of $\mathfrak{a}$, i.e. $X = K a_1 K^{-1}$ for some $a_1 \in \mathfrak{a}$.

Then, any arbitrary element of the group G can be written as

$$G = K_0 \exp(X) = K_0 \exp(Ad_K(a_1)) = K_1 \exp(a_1) K_2, \tag{17}$$

for some $X \in \mathfrak{p}$ where $K_i \in K$ and $a_1 \in \mathfrak{a}$. The first equation is a fact about geodesics in G/K space [24], where $K = \exp(\mathfrak{k})$ is a closed subgroup of G. Eq. (17) is called the KAK decomposition [24].

The results in this chapter suggest that K_1 and K_2 can be synthesized by unbounded controls X_i in negligible time. The time consuming part of the evolution

$\exp(a_1)$ is synthesized by evolution of Hamiltonian X_d. Time optimal strategy suggests evolving X_d and its conjugates $\mathcal{W}_k X_d \mathcal{W}_k^{-1}$ where $\mathcal{W}_k X_d \mathcal{W}_k^{-1}$ all commute.

Written as evolution

$$G = K_1 \prod_k \exp\left(t_k \mathcal{W}_k X_d \mathcal{W}_k^{-1}\right) K_2 = K_1 \prod_k \mathcal{W}_k \exp(t_k X_d) \mathcal{W}_k^{-1} K_2.$$

where $K_1, K_2, \mathcal{W}_k$ take negligible time to synthesize using unbounded controls u_i and time-optimality is characterized by synthesis of commuting Hamiltonians $\mathcal{W}_k X_d \mathcal{W}_k^{-1}$. This characterization of time optimality, involving commuting Hamiltonians is derived using convexity ideas [4, 28]. The remaining chapter develops these notions.

The chapter is organized as follows. In Section 2, we study the $\frac{SU(n)}{SO(n)}$ problem. In Section 3, we study the general $\frac{G}{K}$ problem. The main contribution of the chapter is, we carry out a *global analysis* of time optimality.

Given Lie algebra $\mathfrak{g}$, we use killing form $\langle x, y\rangle = tr(ad_x ad_y)$ as an inner product on $\mathfrak{g}$. When $\mathfrak{g} = su(n)$, we also use the inner product $\langle x, y\rangle = tr(x'y)$. We call this standard inner product.

2. Time optimal control for $SU(n)/SO(n)$ problem

Remark 1. Birkhoff's convexity states, a real $n \times n$ matrix A is doubly stochastic ($\sum_i A_{ij} = \sum_j A_{ij} = 1$, for $A_{ij} \geq 0$) if it can be written as convex hull of permutation matrices P_i (only one 1 and everything else zero in every row and column). Given $\Theta \in SO(n)$ and $X = \begin{bmatrix} \lambda_1 & 0 & \dots & 0 \\ 0 & \lambda_2 & \dots & 0 \\ \vdots & \vdots & \ddots & \vdots \\ 0 & 0 & \dots & \lambda_n \end{bmatrix}$, we have $diag\left(\Theta X \Theta^T\right) = B\ diag(X)$ where diag(X) is a column vector containing diagonal entries of X and $B_{ij} = \left(\Theta_{ij}\right)^2$ and hence $\sum_i B_{ij} = \sum_j B_{ij} = 1$, making B a doubly stochastic matrix, which can be written as convex sum of permutations. Therefore B diag$(X) = \sum_i \alpha_i P_i$ diag(X), i.e. diagonal of a symmetric matrix $\Theta X \Theta^T$, lies in convex hull of its eigenvalues and its permutations. This is called Schur convexity.

Remark 2. $G = SU(n)$ has a closed subgroup $K = SO(n)$ and a Cartan decomposition of its Lie algerbra $\mathfrak{g} = su(n)$ as $\mathfrak{g} = \mathfrak{p} \oplus \mathfrak{k}$, for $\mathfrak{k} = so(n)$ and $p = -iA$ where A is traceless symmetric and $\mathfrak{a}$ is maximal abelian subalgebra of $\mathfrak{p}$, such that $\mathfrak{a} = -i \begin{bmatrix} \lambda_1 & \dots & 0 \\ 0 & \ddots & 0 \\ 0 & 0 & \lambda_n \end{bmatrix}$, where $\sum_i \lambda_i = 0$. **KAK** decomposition in Eq. (17) states for $U \in SU(n)$, $U = \Theta_1 \exp(\Omega) \Theta_2$ where $\Theta_1, \Theta_2 \in SO(n)$ and

$$\Omega = -i \begin{bmatrix} \lambda_1 & \dots & 0 \\ 0 & \ddots & 0 \\ 0 & 0 & \lambda_n \end{bmatrix},$$

where $\sum_i \lambda_i = 0$.

Remark 3. We now give a proof of the reachable set (16), for the $\frac{SU(4)}{SO(4)}$ problem. Let $U(t) \in SU(n)$ be a solution to the differential Eq. (14)

$$\dot{U} = \left(X_d + \sum_i u_i X_i\right)U, \quad U(0) = I.$$

To understand the reachable set of this system we make a change of coordinates $P(t) = K^{'}(t)U(t)$, where, $\dot{K} = (\sum_i u_i X_i)K$. Then

$$\dot{P}(t) = Ad_{K^{'}(t)}(X_d)P(t), \quad Ad_K(X_d) = KXK^{-1}.$$

If we understand reachable set of $P(t)$, then the reachable set in Eq. (14) is easily derived.

Theorem 1. Let $P(t) \in SU(n)$ be a solution to the differential equation

$$\dot{P} = Ad_{K(t)}(X_d)P,$$

and $K(t) \in SO(n)$ and $X_d = -i\begin{bmatrix} \lambda_1 & 0 & \dots & 0 \\ 0 & \lambda_2 & \dots & 0 \\ \vdots & \vdots & \ddots & \vdots \\ 0 & 0 & & \lambda_n \end{bmatrix}$. The elements of the reachable set at time T, take the form $K_1 \exp(-i\mu T)K_2$, where $K_1, K_2 \in SO(n)$ and $\mu \prec \lambda$ (μ lies in convex hull of λ and its permutations), where $\lambda = (\lambda_1, \dots, \lambda_n)^{'}$.

Proof. As a first step, discretize the evolution of $P(t)$, as piecewise constant evolution, over steps of size τ. The total evolution is then

$$P_n = \prod_i \exp(Ad_{k_i}(X_d)\tau), \tag{18}$$

For $t \in [(n-1)\tau, n\tau]$, choose small step Δ, such that $t + \Delta < n\tau$, then $P(t+\Delta) = \exp(Ad_K(X_d)\Delta)P(t)$.

By KAK, $P(t) = K_1 \underbrace{\begin{bmatrix} \exp(i\phi_1) & 0 & 0 & 0 \\ 0 & \exp(i\phi_2) & 0 & 0 \\ 0 & 0 & \ddots & 0 \\ 0 & 0 & 0 & \exp(i\phi_n) \end{bmatrix}}_{A} K_2,$

where $K_1, K_2 \in SO(n)$. To begin with, assume eigenvalues $\phi_j - \phi_k \neq n\pi$, where n is an integer. When we take a small step of size Δ, $P(t)$ changes to $P(t+\Delta)$ as K_1, K_2, A change to

$$K_1(t+\Delta) = \exp(\Omega_1\Delta)K_1, \quad K_2(t+\Delta) = \exp(\Omega_2\Delta)K_2, \quad A(t+\Delta) = \exp(a\Delta)A,$$

where, $\Omega_1, \Omega_2 \in \mathfrak{k}$ and $a \in \mathfrak{a}$. Let $Q(t+\Delta) = K_1(t+\Delta)A(t+\Delta)K_2(t+\Delta)$, which can be written as

$$Q(t+\Delta) = \exp(\Omega_1\Delta)K_1 \exp(a\Delta)A \exp(\Omega_2\Delta)K_2. \tag{19}$$

$$Q(t+\Delta) = \exp(\Omega_1\Delta)\exp\left(K_1 a K_1^{'}\Delta\right)\exp\left(K_1 A \Omega_2 A^{'} K^{'}{}_1 \Delta\right)P(t). \tag{20}$$

Observe

$$P(t+\Delta) = \exp(Ad_K(X_d)\Delta)P(t). \tag{21}$$

We equate $P(t+\Delta)$ and $Q(t+\Delta)$ to first order in Δ. This gives,

$$Ad_K(X_d) = \Omega_1 + K_1 a K_1' + K_1 A \Omega_2 A' K'_1. \tag{22}$$

Multiplying both sides with $K_1'(\cdot)K_1$ gives

$$Ad_{\overline{K}}(X_d) = \Omega_1' + a + A\Omega_2 A'. \tag{23}$$

where, $\overline{K} = K_1' K$ and $\Omega_1' = K'\Omega K$.

We evaluate $A\Omega_2 A^\dagger$, for $\Omega_2 \in so(n)$.

$$\{A\Omega_2 A^\dagger\}_{kl} = \exp\{i(\phi_k - \phi_l)\}(\Omega_2)_{kl} = \underbrace{\cos(\phi_k - \phi_l)(\Omega_2)_{kl}}_{S_{kl}} + i\underbrace{\sin(\phi_k - \phi_l)(\Omega_2)_{kl}}_{R_{kl}}. \tag{24}$$

such that S is skew symmetric and R is traceless symmetric matrix with $iR \in \mathfrak{p}$. Note $iR \perp \mathfrak{a}$ and onto $\mathfrak{a}^\perp$, by appropriate choice of Ω_2.

Given $Ad_{\overline{K}}(X_d) \in \mathfrak{p}$, we decompose it as

$$Ad_{\overline{K}}(X_d) = P\left(Ad_{\overline{K}}(X_d)\right) + Ad_{\overline{K}}(X_d)^\perp = \Omega_1' + a + A\Omega_2 A',$$

with P denoting the projection onto $\mathfrak{a}$ ($\mathfrak{a} = -i\begin{bmatrix} \lambda_1 & \dots & 0 \\ 0 & \ddots & 0 \\ 0 & 0 & \lambda_n \end{bmatrix}$, where $\sum_i \lambda_i = 0$.) w.r.t to standard inner product and $Ad_{\overline{K}}(X_d)^\perp$ to the orthogonal component. In Eq. (24), $\phi_k - \phi_l \neq 0, \pi$, we can solve for $(\Omega_2)_{kl}$ such that $iR = Ad_{\overline{K}}(X_d)^\perp$. This gives Ω_2. Let $a = P\left(Ad_{\overline{K}}(X_d)\right)$ and choose $\Omega_1' = Ad_{\overline{K}}(X_d)^\perp - A\Omega_2 A^\dagger = -S \in \mathfrak{k}$.

With this choice of Ω_1, Ω_2 and a, $P(t+\Delta)$ and $Q(t+\Delta)$ are matched to first order in Δ and

$$P(t+\Delta) - Q(t+\Delta) = o\left(\Delta^2\right).$$

Consider the case, when A is degenerate. Let,

$$A = \begin{bmatrix} A_1 & 0 & \dots & 0 \\ 0 & A_2 & \dots & 0 \\ \vdots & \vdots & \ddots & \vdots \\ 0 & 0 & & A_n \end{bmatrix}, \tag{25}$$

where A_k is n_k fold degenerate (modulo sign) described by $n_k \times n_k$ block. WLOG, we arrange

$$A_k = \exp(i\phi_k)\begin{bmatrix} I_{r\times r} & 0 \\ 0 & -I_{s\times s} \end{bmatrix}. \tag{26}$$

Consider the decomposition

$$Ad_{\overline{K}}(X_d) = P\left(Ad_{\overline{K}}(X_d)\right) + Ad_{\overline{K}}(X_d)^\perp,$$

where P denotes projection onto $n_k \times n_k$ blocks in Eq. (25) and $Ad_K(X_d)^\perp$, the orthogonal complement.

$$P\left(\begin{bmatrix} X_{11} & X_{12} & \dots & X_{1n} \\ X_{21} & X_{22} & \dots & X_{2n} \\ \vdots & \vdots & \ddots & \vdots \\ X_{n1} & X_{n2} & \dots & X_{nn} \end{bmatrix}\right) = \begin{bmatrix} X_{11} & 0 & \dots & 0 \\ 0 & X_{22} & \dots & 0 \\ \vdots & \vdots & \ddots & \vdots \\ 0 & 0 & \dots & X_{nn} \end{bmatrix}, \quad (27)$$

where X_{ij} are blocks.
Then we write

$$Q(t+\Delta) = \exp(\Omega_1\Delta)K_1 \exp\left(P\left(Ad_{\overline{K}}(X_d)\Delta\right)\right)A \exp(\Omega_2\Delta)K_2. \quad (28)$$

where in Eq. (24) we can solve for $(\Omega_2)_{kl}$ such that $iR = Ad_{\overline{K}}(X_d)^{\perp}$. This gives Ω_2. Choose, $Ad_{\overline{K}}(X_d)^{\perp} - A\Omega_2 A^{\dagger} = \Omega_1' \in \mathfrak{k}$, this gives $\Omega_1 = K_1\Omega_1' K_1'$. Again $P(t+\Delta) - Q(t+\Delta) = o(\Delta^2)$. We write Eq. (28) slightly differently.

Let H_1 be a rotation formed from block diagonal matrix

$$H_1 = \begin{bmatrix} \Theta_1 & 0 & \dots & 0 \\ 0 & \Theta_2 & \dots & 0 \\ \vdots & \vdots & \ddots & \vdots \\ 0 & 0 & \dots & \Theta_n \end{bmatrix}, \quad (29)$$

where Θ_k is $n_k \times n_k$ sub-block in $SO(n_k)$. $H_1 = \exp(h_1)$ is chosen such that

$$H_1' P\left(Ad_{\overline{K}}(X_d)\right)H_1 = a$$

is a diagonal matrix. Let $H_2 = \exp(\underbrace{A^{-1}h_1 A}_{h_2})$, where h_2 is skew symmetric, such that

$$h_1 = \begin{bmatrix} \theta_1 & 0 & \dots & 0 \\ 0 & \theta_2 & \dots & 0 \\ \vdots & \vdots & \ddots & \vdots \\ 0 & 0 & \dots & \theta_n \end{bmatrix}, h_2 = \begin{bmatrix} \hat{\theta}_1 & 0 & \dots & 0 \\ 0 & \hat{\theta}_2 & \dots & 0 \\ \vdots & \vdots & \ddots & \vdots \\ 0 & 0 & \dots & \hat{\theta}_n \end{bmatrix}, \quad (30)$$

where
$\theta_k, \hat{\theta}_k$ is $n_k \times n_k$ sub-block in $so(n_k)$, related by (see 26)

$$\hat{\theta}_k = A_k' \theta_k A_k, \quad \theta_k = \begin{bmatrix} \overbrace{\theta_{11}}^{r\times r} & \theta_{12} \\ -\theta_{12}^{\dagger} & \underbrace{\theta_{22}}_{s\times s} \end{bmatrix}, \hat{\theta}_k = \begin{bmatrix} \theta_{11} & -\theta_{12} \\ \theta_{12}^{\dagger} & \theta_{22} \end{bmatrix} \quad (31)$$

Note $H_1'P(Ad_k(X_d))H_1 = a$ lies in convex hull of eigenvalues of X_d. This is true if we look at the diagonal of $H_1' Ad_K(X_d)H_1$, it follows from Schur Convexity. The diagonal of $H_1' Ad_k(X_d)^{\perp}H_1$ is zero as its inner product

$$tr\left(a_1 H_1' Ad_k(X_d)^{\perp}H_1\right) = tr\left(H_1 a_1 H_1' Ad_k(X_d)^{\perp}\right) = 0.$$

as $H_1 a_1 H_1^{'}$ has block diagonal form which is perpendicular to $Ad_k(X_d)^{\perp}$. Therefore diagonal of $H_1^{'}P(Ad_k(X_d))H_1$ is same as diagonal of $H_1^{'}Ad_K(X_d)H_1$.

Now using $H_1 A H_2^{\dagger} = A$, from 28, we have

$$Q(t+\Delta) = \exp(\Omega_1\Delta)K_1\exp\left(P\left(Ad_{\overline{K}}(X_d)\Delta\right)\right)H_1 A H_2^{\dagger}\exp(\Omega_2\Delta)K_2. \quad (32)$$

$$Q(t+\Delta) = \exp(\Omega_1\Delta)K_1 H_1\exp(a\Delta)AH_2^{\dagger}\exp(\Omega_2\Delta)K_2. \quad (33)$$

where the above expression can be written as

$$Q(t+\Delta) = \exp(\Omega_1\Delta)\exp\left(K_1 H_1 a H_1^{'}K_1^{'}\Delta\right)\exp\left(K_1 A\Omega_2 A^{'}K^{'}{}_1\Delta\right)P(t).$$

where Ω_1, H_1, a, Ω_2, are chosen such that

$$\left(\Omega_1 + K_1 H_1 a H_1^{'}K_1^{'} + K_1 A\Omega_2 A^{'}K^{'}{}_1\right) = Ad_K(X_d).$$

$$\left(\Omega_1^{'} + H_1 a H_1^{'} + A\Omega_2 A^{'}\right) = Ad_{\overline{K}}(X_d).$$

$$Q(t+\Delta) - P(t+\Delta) = o\left(\Delta^2\right)P(t).$$

$$Q(t+\Delta) = \left(I + o\left(\Delta^2\right)\right)P(t+\Delta).$$

$$\begin{aligned} Q(t+\Delta)Q(t+\Delta)^T &= \left(I+o\left(\Delta^2\right)\right)P(t+\Delta)P^T(t+\Delta)\left(I+o\left(\Delta^2\right)\right) \\ &= P(t+\Delta)P^T(t+\Delta)\left[I+o\left(\Delta^2\right)\right]. \end{aligned}$$

$$P(t+\Delta)P^T(t+\Delta) = K_1\begin{bmatrix} \exp(i2\phi_1) & 0 & \dots & 0 \\ 0 & \exp(i2\phi_2) & \dots & 0 \\ \vdots & \vdots & \ddots & \vdots \\ 0 & 0 & \dots & \exp(i2\phi_n) \end{bmatrix}K_1^T.$$

Let $F = P(t+\Delta)P^T(t+\Delta)$ and $G = Q(t+\Delta)Q^T(T+\Delta)$ we relate the eigenvalues, of F and G. Given F, G, as above, with $|F-G| \le \varepsilon$, and a ordered set of eigenvalues of F, denote $\lambda(F) = \begin{bmatrix} \exp(i2\phi_1) \\ \exp(i2\phi_2) \\ \vdots \\ \exp(i2\phi_n) \end{bmatrix}$, there exists an ordering (correspondence) of eigenvalues of G, such that $|\lambda(F) - \lambda(G)| < \varepsilon$.

Choose an ordering of $\lambda(G)$ call μ that minimizes $|\lambda(F) - \lambda(G)|$.

$F = U_1 D(\lambda)U_1^{'}$ and $G = U_2 D(\mu)U_{2^{'}}$, where $D(\lambda)$ is diagonal with diagonal as λ, let $U = U_1^{'}U_2$,

$$|F-G|^2 = \left|D(\lambda) - UD(\mu)U^{'}\right|^2 = |\lambda|^2 + |\mu|^2 - tr\left(D(\lambda)^{'}UD(\mu)U^{'} + (UD(\mu)U)^{'}D(\lambda)\right),$$

By Schur convexity,

$$tr\left(D(\lambda)^{'}UD(\mu)U^{'} + \left(UD(\mu)U^{'}\right)^{'}D(\lambda)\right) = \sum_i \alpha_i\left(\lambda^{'}P_i(\mu) + P_i(\mu)^{'}\lambda\right),$$

where P_i are permutations. Therefore $|F-G|^2 > |\lambda - \mu|^2$.
Therefore,

$$\lambda\left(QQ^T(t+\Delta)\right) = \lambda\left(PP^T(t+\Delta)\right) + o\left(\Delta^2\right).$$

The difference

$$o\left(\Delta^2\right) = \underbrace{\exp\left(\left(\Omega_1 + K_1 H_1 a H_1^{'} K_1^{'} + K_1 A \Omega_2 A^{'} K^{'}{}_1\right)\Delta\right)}_{\exp\left(Ad_K(X_d)\Delta\right)}$$
$$- \exp\left(\Omega_1 \Delta\right) \exp\left(K_1 H_1 a H_1^{'} K_1^{'} \Delta\right) \exp\left(K_1 A \Omega_2 A^{'} K^{'}{}_1 \Delta\right),$$

is regulated by size of Ω_2, which is bounded by $|\Omega_2| \leq \frac{\|X_d\|}{\sin\left(\phi_i - \phi_j\right)}$, where $\sin\left(\phi_i - \phi_j\right)$ is smallest non-zero difference. Δ is chosen small enough such that $|o\left(\Delta^2\right)| < \varepsilon\Delta$.

For each point $t \in [0, T]$, we choose an open nghd $N(t) = (t - N_t, t + N_t)$, such that $o_t\left(\Delta^2\right) < \varepsilon\Delta$ for $\Delta \in N(t)$. $N(t)$ forms a cover of $[0, T]$. We can choose a finite subcover centered at $t_1, \ldots, t_n$ (see **Figure 1A**). Consider trajectory at points $P(t_1), \ldots, \ldots P(t_n)$. Let $t_{i,i+1}$ be the point in intersection of $N(t_i)$ and $N(t_{i+1})$. Let $\Delta_i^+ = t_{i,i+1} - t_i$ and $\Delta_{i+1}^- = t_{i+1} - t_{i,i+1}$. We consider points $P(t_i), P(t_{i+1}), P(t_{i,i+1}), \underbrace{Q\left(t_i + \Delta_i^+\right)}_{Q_{i+}}, \underbrace{Q\left(t_{i+1} - \Delta_{i+1}^-\right)}_{Q_{(i+1)-}}$ as shown in **Figure 1B**.

Then we get the following recursive relations.

$$\lambda\left(Q_{i+} Q_{i+}^T\right) = \exp\left(2a_i^+ \Delta_i^+\right)\ \lambda\left(P_i P_i^T\right) \tag{34}$$

$$\lambda\left(P_{i,i+1} P_{i,i+1}^T\right) = \lambda\left(Q_{i+} Q_{i+}^T\right) + o\left(\left(\Delta_i^+\right)^2\right) \tag{35}$$

$$\lambda\left(Q_{(i+1)-} Q_{(i+1)-}^T\right) = \lambda\left(P_{i,i+1} P_{i,i+1}^T\right) + o\left(\left(\Delta_{i+1}^-\right)^2\right) \tag{36}$$

$$\exp\left(-2a_{i+1}^- \Delta_{i+1}^-\right)\ \lambda\left(P_{i+1} P_{i+1}^T\right) = \lambda\left(Q_{(i+1)-} Q_{(i+1)-}^T\right) \tag{37}$$

where a_i^+ and a_{i+1}^- correspond to a in Eq. (33) and lie in the convex hull of the eigenvalues X_d.

Adding the above equations,

$$\lambda\left(P_{i+1} P_{i+1}^T\right) = \exp\left(o\left(\Delta^2\right)\right) \exp\left(2\left(a_i^+ \Delta_i^+ + a_{i+1}^- \Delta_{i+1}^-\right)\ \lambda\left(P_i P_i^\dagger\right)\right). \tag{38}$$

$$\lambda\left(P_n P_n^T\right) = \exp\left(\underbrace{\sum o\left(\Delta^2\right)}_{\leq \varepsilon T}\right) \exp\left(2 \sum_i a_i^+ \Delta_i^+ + a_{i+1}^- \Delta_{i+1}^-\right)\ \lambda\left(P_1 P_1^T\right). \tag{39}$$

where $o\left(\Delta^2\right)$ in Eq. (38) is diagonal.

$$\lambda\left(P_n P_n^T\right) = \exp\left(\underbrace{\sum o\left(\Delta^2\right)}_{\leq \varepsilon T}\right)\ \exp\left(2T \sum_k \alpha_k P_k(\lambda)\right)\ \lambda\left(P_1 P_1^T\right) = \exp\left(\underbrace{\sum o\left(\Delta^2\right)}_{\leq \varepsilon T}\right) \exp\left(2\mu T\right)\ \lambda\left(P_1 P_1^T\right), \tag{40}$$

where $\mu < \lambda$ and $P_1 = I$.

$$P_n = K_1 \exp\left(\frac{1}{2} \underbrace{\sum o\left(\Delta^2\right)}_{\leq \varepsilon T}\right)\ \exp\left(\mu T\right)\ K_2. \tag{41}$$

Note, $|P_n - K_1 \exp\left(\mu T\right) K_2| = o(\varepsilon)$. This implies that P_n belongs to the compact set $K_1 \exp\left(\mu T\right) K_2$, else it has minimum distance from this compact set and by making $\Delta \to 0$ and hence $\varepsilon \to 0$, we can make this arbitrarily small. In Eq. (18), $P_n \to P(T)$ as $\tau \to 0$. Hence $P(T)$ belongs to compact set $K_1 \exp(\mu T) K_2$. **q.e.d.**

Corollary 1. Let $U(t) \in SU(n)$ be a solution to the differential equation

$$\dot{U} = \left(X_d + \sum_i u_i X_i \right) U,$$

where $\{X_i\}_{LA}$, the Lie algebra generated by X_i, is $so(n)$ and $X_d = -i \begin{bmatrix} \lambda_1 & 0 & \dots & 0 \\ 0 & \lambda_2 & \dots & 0 \\ \vdots & \vdots & \ddots & \vdots \\ 0 & 0 & \dots & \lambda_n \end{bmatrix}$. The elements of reachable set at time T, takes the form $U(T) \in K_1 \exp(-i\mu T) K_2$, where $K_1, K_2 \in SO(n)$ and $\mu \prec \lambda$, where $\lambda = (\lambda_1, \dots, \lambda_n)'$ and the set $S = K_1 \exp(-i\mu T) K_2$ belongs to the closure of reachable set.

Proof. Let $V(t) = K'(t) U(t)$, where, $\dot{K} = \left(\sum_i u_i X_i\right) K$. Then

$$\dot{V}(t) = Ad_{K'(t)}(X_d) V(t).$$

From Theorem 1, we have $V(T) \in K_1 \exp(-i\mu T) K_2$. Therefore $U(T) \in K_1 \exp(-i\mu T) K_2$. Given

$$U = K_1 \exp(-i\mu T) K_2 = K_1 \exp\left(-i \sum_j \alpha_j P_j(\lambda) T \right) K_2$$
$$= K_1 \prod_j \exp(-i t_j X_d) K_j, \quad \sum t_j = T.$$

We can synthesize K_j in negligible time, therefore $|U(T) - U| < \varepsilon$, for any desired ε. Hence U is in closure of reachable set. **q.e.d.**

Remark 4. We now show how Remark 2 and Theorem 1 can be mapped to results on decomposition and reachable set for coupled spins/qubits. Consider the transformation

$$W = \exp(-i\pi I_y S_y) \exp\left(-i\frac{\pi}{2} I_z\right)$$

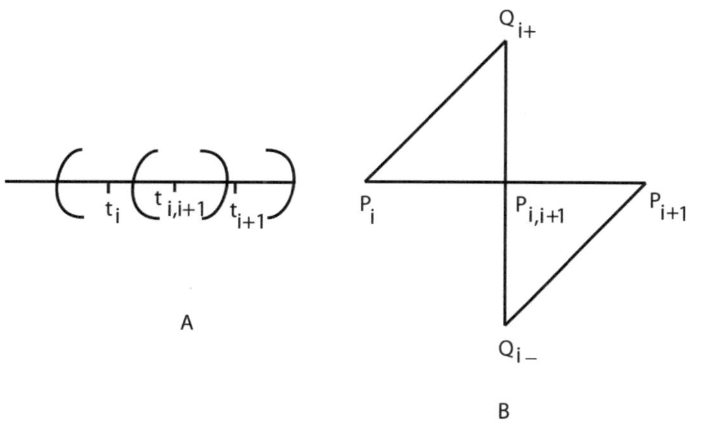

Figure 1.
A. Collection of overlapping neighborhoods forming the finite subcover. B. Depiction of P_i, P_{i+1}, Q_{i+}, Q_{i-}, $P_{i,i+1}$ as in proof of Theorem 1.

The transformation maps the algebra $\mathfrak{k} = su(2) \times su(2) = \{I_\alpha, S_\alpha\}$ to $\mathfrak{k}_1 = so(4)$, four dimensional skew symmetric matrices, i.e., $Ad_W(\mathfrak{k}) = \mathfrak{k}_1$. The transformation maps $\mathfrak{p} = \{I_\alpha S_\beta\}$ to $\mathfrak{p}_1 = -iA$, where A is traceless symmetric and maps $\mathfrak{a} = -i\{I_x S_x, I_y S_y, I_z S_z\}$ to $\mathfrak{a}_1 = -i\{-\frac{S_z}{2}, \frac{I_z}{2}, I_z S_z\}$, space of diagonal matrices in $\mathfrak{p}_1$, such that $a_x I_x S_x + a_y I_y S_y + a_z I_z S_z$ gets mapped to the four vector (the diagonal) $(\lambda_1, \lambda_2, \lambda_3, \lambda_4) = (a_y + a_z - a_x, a_x + a_y - a_z, -(a_x + a_y + a_z), a_x + a_z - a_y)$.

Corollary 2. Canonical decomposition. Given the decomposition of SU(4) from Remark 2, we can write

$$U = \exp(\Omega_1) \exp\left(-i \begin{bmatrix} \lambda_1 & \dots & 0 \\ 0 & \ddots & 0 \\ 0 & 0 & \lambda_4 \end{bmatrix}\right) \exp(\Omega_2),$$

where $\Omega_1, \Omega_2 \in so(4)$. We write above as

$$U = \exp(\Omega_1) \exp\left(-i\left(-\frac{a_x}{2} S_z + \frac{a_y}{2} I_z + a_z I_z S_z\right)\right) \exp(\Omega_2),$$

Multiplying both sides with $W^{'}(.)W$ gives

$$W^{'} U W = K_1 \exp(-i a_x I_x S_x + a_y I_y S_y + a_z I_z S_z) K_2,$$

where $K_1, K_2 \in SU(2) \times SU(2)$ local unitaries and we can rotate to $a_x \geq a_y \geq |a_z|$.

Corollary 3. Digonalization. Given $-iH_c = -i\sum_{\alpha\beta} J_{\alpha\beta} I_\alpha S_\beta$, there exists a local unitary K such that

$$K(-iH_c)K^{'} = -i(a_x I_x S_x + a_y I_y S_y + a_z I_z S_z), \; a_x \geq a_y \geq |a_z|.$$

Note $W(-iH_c)W^{'} \in \mathfrak{p}_1$. Then choose $\Theta \in SO(n)$ such that $\Theta W(-iH_c)W^{'}\Theta^{'} = -i(-\frac{a_x}{2} S_z + \frac{a_y}{2} I_z + a_z I_z S_z)$ and hence

$$\left(W^{'} \exp(\Omega) W\right)(-iH_c)\left(W \exp(\Omega) W^{'}\right)^{'} = -i(a_x I_x S_x + a_y I_y S_y + a_z I_z S_z).$$

where $K = W^{'} \exp(\Omega) W$ is a local unitary. We can rotate to ensure $a_x \geq a_y \geq |a_z|$.

Corollary 4. Given the evolution of coupled qubits $\dot{U} = -i\left(H_c + \sum_j u_j H_j\right)U$, we can diagonalize $H_c = \sum_{\alpha\beta} J_{\alpha\beta} I_\alpha S_\beta$ by local unitary $X_d = K^{'} H_c K = a_x I_x S_x + a_y I_y S_y + a_z I_z S_z$, $a_x \geq a_y \geq |a_z|$, which we write as triple (a_x, a_y, a_z). From this, there are 24 triples obtained by permuting and changing sign of any two by local unitary. Then $U(T) \in S$ where

$$S = K_1 \exp\left(T \sum_i \alpha_i (a_i, b_i, c_i)\right) K_2, \quad \alpha_i > 0 \quad \sum_i \alpha_i = 1.$$

Furthermore S belongs to the closure of the reachable set. Alternate description of S is

$$U = K_1 \exp(-i(\alpha I_x S_x + \beta I_y S_y + \gamma I_z S_z)) K_2, \quad \alpha \geq \beta \geq |\gamma|,$$

$\alpha \leq a_x T$ and $\alpha + \beta \pm \gamma \leq (a_x + a_y \pm a_z)T$.

Proof. Let $V(t) = K^{'}(t)U(t)$, where $\dot{K} = \left(-i\sum_j u_j X_j\right)K$. Then

$$\dot{V}(t) = Ad_{K'(t)}(-iX_d)V(t).$$

Consider the product

$$V = \prod_i \exp\left(Ad_{K_i}(-iX_d)\Delta t\right)$$

where $K_i \in SU(2) \otimes SU(2)$ and $X_d = a_x I_x S_x + a_y I_y S_y + a_z I_z S_z$, where $a_x \geq a_y \geq |a_z|$. Then,

$$WVW' = \prod_i \exp\left(Ad_{WK_iW'}(-iWX_dW')\Delta t\right)$$

Observe $WK_iW' \in SO(4)$ and $WX_dW' = \mathrm{diag}(\lambda_1, \lambda_2, ..., \lambda_4)$. Then using results from Theorem 1, we have

$$WVW' = J_1 \exp(-i\mu) J_2 = J_1 \exp\left(-i \sum_j \alpha_j P_j(\lambda)\right) J_2, \quad J_1, J_2 \in SO(4), \quad \mu < \lambda T$$

Multiplying both sides with $W'(\cdot)W$, we get

$$V = K_1 \exp\left(T \sum_i \alpha_i(a_i, b_i, c_i)\right) K_2, \quad \alpha_i > 0 \quad \sum_i \alpha_i = 1.$$

which we can write as

$$V = K_1 \exp\left(-i(\alpha I_x S_x + \beta I_y S_y + \gamma I_z S_z)\right) K_2, \quad \alpha \geq \beta \geq |\gamma|,$$

where using $\mu < \lambda T$, we get,

$$\alpha + \beta - \gamma \leq (a_x + a_y - a_z)T \tag{42}$$

$$\alpha \leq a_x T \tag{43}$$

$$\alpha + \beta + \gamma \leq (a_x + a_y + a_z)T. \tag{44}$$

Furthermore $U = KV$. Hence the proof. **q.e.d.**

3. Time optimal control for G/K problem

Remark 5. Stabilizer: Let $\mathfrak{g} = \mathfrak{p} \oplus \mathfrak{k}$ be Cartan decomposition of real semisimple Lie algebra $\mathfrak{g}$ and $\mathfrak{a} \in \mathfrak{p}$ be its Cartan subalgebra. Let $a \in \mathfrak{a}$. $ad_a^2 : \mathfrak{p} \to \mathfrak{p}$ is symmetric in basis orthonormal wrt to the killing form. We can diagonalize ad_a^2. Let Y_i be eigenvectors with nonzero (negative) eigenvalues $-\lambda_i^2$. Let $X_i = \frac{[a, Y_i]}{\lambda_i}$, $\lambda_i > 0$.

$$ad_a(Y_i) = \lambda_i X_i, \quad ad_a(X_i) = -\lambda_i Y_i.$$

X_i are independent, as $\sum \alpha_i X_i = 0$ implies $-\sum \alpha_i \lambda_i Y_i = 0$. Since Y_i are independent, X_i are independent. Given $X \perp X_i$, then $[a, X] = 0$, otherwise we can decompose it in eigenvectors of ad_a^2, i.e., $[a, X] = \sum_i \alpha_i a_i + \sum_j \beta_j Y_j$, where a_i are zero eigenvectors of ad_a^2. Since $0 = \langle X[a[a, X]] \rangle = -\|[a, X]\|^2$, which means $[a, X] = 0$. This is a contradiction. Y_i are orthogonal, implies X_i are orthogonal,

$\langle[a, Y_i][a, Y_j]\rangle = \langle[a, [a, Y_i]Y_j\rangle = \lambda_i^2\langle Y_i Y_j\rangle = 0$. Let $\mathfrak{k}_0 \in \mathfrak{k}$ satisfy $[a, \mathfrak{k}_0] = 0$. Then $\mathfrak{k}_0 = \{X_i\}^{\perp}$.

$\tilde{Y}_i$ denote eigenvectors that have λ_i as non-zero integral multiples of π. $\tilde{X}_i$ are ad_a related to $\tilde{Y}_i$. We now reserve Y_i for non-zero eigenvectors that are not integral multiples of π.

Let

$$\mathfrak{f} = \{a_i\} \oplus \tilde{Y}_i, \qquad \mathfrak{h} = \mathfrak{k}_0 \oplus \tilde{X}_i,$$

$\tilde{X}_i, X_l, k_j$ where k_j forms a basis of $\mathfrak{k}_0$, forms a basis of $\mathfrak{k}$. Let $A = \exp(a)$.
$AkA^{-1} = A\left(\sum_i \alpha_i X_i + \sum_l \alpha_l \tilde{X}_l + \sum_j \alpha_j k_j\right)A^{-}$, where $k \in \mathfrak{k}$

$$AkA^{-1} = \textstyle\sum_i \alpha_i[\cos(\lambda_i)X_i - \sin(\lambda_i)Y_i] + \sum_l \pm \alpha_l \tilde{X}_l + \sum_j \alpha_j k_j$$

The range of $A(\cdot)A^{-1}$ in $\mathfrak{p}$, is perpendicular to $\mathfrak{f}$. Given $Y \in \mathfrak{p}$ such that $Y \in \mathfrak{f}^{\perp}$. The norm $\|X\|$ of $X \in \mathfrak{k}$, such that $\mathfrak{p}$ part of $AXA^{-1}|_{\mathfrak{p}} = Y$ satisfies

$$\|X\| \leq \frac{\|Y\|}{\sin\lambda_s}. \tag{45}$$

where λ_s^2 is the smallest nonzero eigenvalue of $-ad_a^2$ such that λ_s is not an integral multiple of π.

A^2kA^{-2} stabilizes $\mathfrak{h} \in \mathfrak{k}$ and $\mathfrak{f} \in \mathfrak{p}$. If $k \in \mathfrak{k}$, is stabilized by $A^2(\cdot)A^{-2}$, $\lambda_i = n\pi$, i.e., $k \in \mathfrak{h}$. This means $\mathfrak{h}$ is an subalgebra, as the Lie bracket of $[y, z] \in \mathfrak{k}$ for $y, z \in \mathfrak{h}$ is stabilized by $A^2(\cdot)A^{-2}$.

Let $H = \exp(\mathfrak{h})$, be an integral manifold of $\mathfrak{h}$. Let $\tilde{H} \in K$ be the solution to $A^2\tilde{H}A^{-2} = \tilde{H}$ or $A^2\tilde{H} - \tilde{H}A^{-2} = 0$. $\tilde{H}$ is closed, $H \in \tilde{H}$. We show that $\tilde{H}$ is a manifold. Given element $H_0 \in \tilde{H} \in K$, where K is closed, we have a $\exp(B_\delta^{\mathfrak{k}})$ nghd of H_0, in $\exp(B_\delta)$ ball nghd of H_0, which is one to one. For $x \in B_\delta^{\mathfrak{k}}$,
$A^2\exp(x)A^{-2} = \exp(x)$, implies,

$$A^2\exp\left(\sum_i \alpha_i X_i + \sum_l \beta_l \tilde{X}_l + \sum_j \gamma_j k_j\right)H_0 A^{-2} = \exp\left(\sum_i \alpha_i \cos(2\lambda_i)X_i - \sin(2\lambda_i)Y_i\right.$$
$$\left. + \sum_l \beta_l \tilde{X}_l + \sum_j \gamma_j k_j\right)H_0 = \exp\left(\sum_i \alpha_i X_i + \sum_l \beta_l \tilde{X}_l + \sum_j \gamma_j k_j\right)H_0,$$

then by one to one property of $\exp(B_\delta)$, we get $\alpha_i = 0$ and $x \in \mathfrak{h}$. Therefore $\exp\left(B_\delta^{\mathfrak{h}}\right)H_0$ is a nghd of H_0.

Given a sequence $H_i \in \exp(\mathfrak{h})$ converging to H_0, for n large enough $H_n \in \exp\left(B_\delta^{\mathfrak{h}}\right)H_0$. Then H_0 is in invariant manifold $\exp(\mathfrak{h})$. Hence $\exp(\mathfrak{h})$ is closed and hence compact.

Let $y \in \mathfrak{f}$, then there exists a $h_0 \in \mathfrak{h}$ such that $\exp(h_0)y\exp(-h_0) \in \mathfrak{a}$. We maximize the function $\langle a_r, \exp(h)y\exp(h)\rangle$, over the compact group $\exp(\mathfrak{h})$, for regular element $a_r \in \mathfrak{a}$ and $\langle ., .\rangle$ is the killing form. At the maxima, we have at $t = 0$,
$\frac{d}{dt}\langle a_r, \exp(h_1 t)(\exp(h_0)y\exp(-h_0))\exp(-h_1 t)\rangle = 0$.

$$\langle a_r, [h_1\exp(h_0)y\exp(-h_0)]\rangle = -\langle h_1, [a_r\exp(h_0)y\exp(-h_0)]\rangle,$$

if $\exp(h_0)y\exp(-h_0) \neq \mathfrak{a}$, then $[a_r, \exp(h_0)y\exp(-h_0)] \in \mathfrak{k}$. The bracket $[a_r, \exp(h_0)y\exp(-h_0)]$ is Ad_{A^2} invariant and, hence, belongs to $\mathfrak{h}$. We can choose h_1 so that gradient is not zero. Hence $\exp(h_0)y\exp(-h_0) \in \mathfrak{a}$. For $z \in \mathfrak{z}$ such that $z \in \mathfrak{f}^{\perp}$, we have $\exp(h_0)z\exp(-h_0) \in \mathfrak{a}^{\perp}$.

$$\langle \mathfrak{a}, \exp(h_0)z\exp(-h_0)\rangle = \langle \exp(-h_0)\mathfrak{a}\exp(h_0), z\rangle = 0,$$

as $\exp(-h_0)\mathfrak{a}\exp(h_0)$ is Ad_{A^2} invariant, hence $\exp(-h_0)\mathfrak{a}\exp(h_0) \in \mathfrak{f}$. In above, we worked with killing form. For $\mathfrak{g} = su(n)$, we may use standard inner product.

Remark 6. Kostant's convexity: [28] Given the decomposition $\mathfrak{g} = \mathfrak{p} \oplus \mathfrak{k}$, let $\mathfrak{a} \subset \mathfrak{p}$ and $X \in \mathfrak{a}$,. Let $\mathcal{W}_i \in \exp(\mathfrak{k})$ such that $\mathcal{W}_i X \mathcal{W}_i \in \mathfrak{a}$ are distinct, Weyl points. Then projection (w.r.t killing form) of $Ad_K(X)$ on $\mathfrak{a}$ lies in convex hull of these Weyl points. The C be the convex hull and let projection $P(Ad_K(X))$ lie outside this Hull. Then there is a separating hyperplane a, such that $\langle Ad_K(X), a\rangle < \langle C, a\rangle$. W.L.O.G we can take a to be a regular element. We minimize $\langle Ad_K(X), a\rangle$, with choice of K and find that minimum happens when $[Ad_K(X), a] = 0$, i.e. $Ad_K(X)$ is a Weyl point. Hence $P(Ad_K(X)) \in \sum_i \alpha_i \mathcal{W}_i X \mathcal{W}_i^{-1}$, for $\alpha_i > 0$ and $\sum_i \alpha_i = 1$. The result is true with a projection w.r.t inner product that satisfies $\langle x, [y, z]\rangle = \langle [x, y], z]\rangle$, like standard inner product on $\mathfrak{g} = su(n)$.

Theorem 2 Given a compact Lie group G and Lie algebra $\mathfrak{g}$. Consider the Cartan decomposition of a real semisimple Lie algebra $\mathfrak{g} = \mathfrak{p} \oplus \mathfrak{k}$. Given the control system

$$\dot{X} = Ad_{K(t)}(X_d)X, \quad P(0) = \mathbf{1}$$

where $X_d \in \mathfrak{a}$, the Cartan subalgebra $\mathfrak{a} \in \mathfrak{p}$ and $K(t) \in \exp \mathfrak{k}$, a closed subgroup of G. The end point

$$P(T) = K_1 \exp\left(T \sum_i \alpha_i \mathcal{W}_i(X_d)\right) K_2,$$

where $K_1, K_2 \in \exp(\mathfrak{k})$ and $\mathcal{W}_i(X_d) \in \mathfrak{a}$ are Weyl points, $\alpha_i > 0$ and $\sum_i \alpha_i = 1$.

Proof. As in proof of Theorem 1, we define

$$P(t+\Delta) = \exp(Ad_K(X_d)\Delta)P(t) = \exp(Ad_K(X_d)\Delta)K_1\exp(a)K_2$$

and show that

$$\exp(Ad_K(X_d)\Delta)K_1AK_2 = K_a\exp(a_0\Delta + C\Delta^2)AK_b = K_a\exp(a + a_0\Delta + C\Delta^2)K_b, \quad (46)$$

where for $\overline{K} = K_1^{-1}K$,

$$Ad_{\overline{K}}(X_d) = \underbrace{P\left(Ad_{\overline{K}}(X_d)\right)}_{a_0} + Ad_{\overline{K}}(X_d)^{\perp}.$$

where P is projection w.r.t killing form and $a_0 \in \mathfrak{f}$, the centralizer in $\mathfrak{p}$ as defined in Remark 5, $C\Delta^2 \in \mathfrak{f}$ is a second order term that can be made small by choosing Δ. $K_a, K_b \in \exp(\mathfrak{k})$.

To show Eq. (46), we show there exists $K_{1'}^{'}, K_{2''} \in K$ such that

$$\underbrace{\exp\left(k_1''\right)}_{K_1''}\ \exp\left(Ad_{\overline{K}}(X_d)\Delta\right)\underbrace{\exp\left(Ak_2''A^{-1}\right)}_{K_2''} = \exp\left(a_0\Delta + C\Delta^2\right), \tag{47}$$

where K_1'' and K_2'' are constructed by a iterative procedure as described in the proof below.

Given X and Y as $N \times N$ matrices, considered elements of a matrix Lie algebra $\mathfrak{g}$, we have,

$$\log\left(e^X e^Y\right) - (X+Y) = \sum_{n>0}\frac{(-1)^{n-1}}{n}\sum_{1\le i\le n}\frac{\left[X^{r_1}Y^{s_1}\ldots X^{r_n}Y^{s_n}\right]}{\sum_{i=1}^{n}(r_i+s_i)r_1!s_1!\ldots r_n!s_n!}, \tag{48}$$

where $r_i + s_i > 0$.

We bound the largest element (absolute value) of $\log\left(e^X e^Y\right) - (X+Y)$, denoted as $\left|\log\left(e^X e^Y\right) - (X+Y)\right|_0$, given $|X|_0 < \Delta$ and $|Y|_0 < b_0\Delta^k$, where $k \ge 1$, $\Delta < 1$, $b_0\Delta < 1$.

$$\left|\log\left(e^X e^Y\right) - (X+Y)\right|_0 \le \sum_{n=1} N b_0 e \Delta^{k+1} + \sum_{n>1}\frac{1}{n}\frac{\left(2Ne^2\right)^n b_0 \Delta^{n+k-1}}{n} \tag{49}$$

$$\le N b_0 e \Delta^{k+1} + \left(Ne^2\right)^2 b_0 \Delta^{k+1}\left(1 + 2Ne^2\Delta + \ldots\right) \tag{50}$$

$$\le N b_0 e \Delta^{k+1} + \frac{\left(Ne^2\right)^2 b_0 \Delta^{k+1}}{1 - 2Ne^2\Delta} \le \tilde{M} b_0 \Delta^{k+1} \tag{51}$$

where $2N\Delta < 1$ and $\tilde{M}\Delta < 1$.

Given decomposition of $\mathfrak{g} = \mathfrak{p} \oplus \mathfrak{k}$, $\mathfrak{p} \perp \mathfrak{k}$ with respect to the negative definite killing form $B(X,Y) = tr(ad_X ad_Y)$. Furthermore there is decomposition of $\mathfrak{p} = \mathfrak{a} \oplus \mathfrak{a}^{\perp}$.

Given

$$U_0 = \exp\left(a_0\Delta + b_0\Delta + c_0\Delta\right),$$

where $a_0 \in \mathfrak{a}$, $b_0 \in \mathfrak{a}^{\perp}$ and $c_0 \in \mathfrak{k}$, such that $|a_0|_0 + |b_0|_0 + |c_0|_0 < 1$, which we just abbreviate as $a_0 + b_0 + c_0 < 1$ (we follow this convention below).

We describe an iterative procedure

$$U_n = \Pi_{k=1}^{n}\exp\left(-c_k\Delta\right)\ U_0\ \Pi_{k=0}^{n}\exp\left(-b_k\Delta\right), \tag{52}$$

where $c_k \in \mathfrak{k}$ and $b_k \in \mathfrak{a}^{\perp}$, such that the limit

$$n \to \infty \quad U_n = \exp\left(a_0\Delta + C\Delta^2\right), \tag{53}$$

where $a_0, C \in \mathfrak{a}$.

$$\begin{aligned} U_1 &= \exp\left(-c_0\Delta\right)\exp\left(a_0\Delta + b_0\Delta + c_0\Delta\right)\exp\left(-b_0\Delta\right) \\ &= \exp\left(a_0\Delta + b_0\Delta + c_{0'}\Delta^2\right)\exp\left(-b_0\Delta\right) \\ &= \exp\left(a_0\Delta + b_{0'}\Delta^2 + c_{0'}\Delta^2\right) \\ &= \exp\left((a_1 + b_1 + c_1)\Delta\right) \end{aligned}$$

Note b_0' and c_0' are elements of $\mathfrak{g}$ and need not be contained in $\mathfrak{a}^{\perp}$ and $\mathfrak{k}$.

Where, using bound in $c_0' \leq \tilde{M}c_0$, which gives $a_0 + b_0 + c_0'\Delta \leq a_0 - b_0 + c_0$. Using the bound again, we obtain, $b_0' \leq \tilde{M}b_0$. We can decompose, $(b_0' + c_0')\Delta$, into subspaces $a_0'' + b_1 + c_1$, where $a_0'' \leq M(b_0' + c_0')\Delta$, $b_1 \leq M(b_0' + c_0')\Delta$ and $c_1 \leq M(b_0' + c_0')\Delta$, where $-B(X,X) \leq \lambda_{max}|X|^2$, where $|X|$ is Frobenius norm and $-B(X,X) \geq \lambda_{min}|X|^2$. Let $M = \frac{N\lambda_{max}}{\lambda_{min}}$.

This gives, $a_0'' \leq M(b_0' + c_0')\Delta$, $b_1 \leq M(b_0' + c_0')\Delta$ and $c_1 \leq M(b_0' + c_0')\Delta$. This gives

$$a_1 \leq a_0 + \tilde{M}M(b_0 + c_0)\Delta \; b_1 \leq \tilde{M}M(b_0 + c_0)\Delta \; c_1 \leq \tilde{M}M(b_0 + c_0)\Delta$$

For $4\tilde{M}M\Delta<1$, we have, $a_1 + b_1 + c_1 \leq a_0 + b_0 + c_0$. Continuing and using $(b_k + c_k) \leq 2\tilde{M}M\Delta(b_{k-1} + c_{k-1}) \leq \left(2\tilde{M}M\Delta\right)^k (b_0 + c_0)$.

Similarly,

$$|a_k - a_{k-1}|_0 \leq \left(2\tilde{M}M\Delta\right)^k (b_0 + c_0)$$

Note, (a_k, b_k, c_k) is a Cauchy sequences which converges to $(a_\infty, 0, 0)$, where

$$|a_\infty - a_0|_0 \leq (b_0 + c_0) \sum_{k=1}^{\infty} \left(2\tilde{M}M\Delta\right)^k \leq \frac{2M\tilde{M}\Delta(b_0 + c_0)}{1 - 2\tilde{M}M\Delta} \leq C\Delta,$$

where $C = 4\tilde{M}M(b_0 + c_0)$.

The above exercise was illustrative. Now we use an iterative procedure as above to show Eq. (47).

Writing

$$Ad_{\overline{K}}(X_d) = \underbrace{P\left(Ad_{\overline{K}}(X_d)\right)}_{a_0} + \underbrace{Ad_{\overline{K}}(X_d)^\perp}_{b_0},$$

where $a_0 \in \mathfrak{f}$ and $b_0 \in \mathfrak{f}^\perp$, consider again the iterations

$$\begin{aligned} U_0 &= \exp(-\bar{c}_0\Delta)\exp(a_0\Delta + b_0\Delta)\exp(-b_0\Delta + \bar{c}_0\Delta) \\ &= \exp(-\bar{c}_0\Delta)\exp\left(a_0\Delta + \bar{c}_0\Delta + b_{0'}\Delta^2\right) \\ &= \exp\left(a_0\Delta + b_{0'}\Delta^2 + c_{0'}\Delta^2\right) \\ &= \exp(a_1\Delta + b_1\Delta + c_1\Delta) \end{aligned}$$

We refer to Remark 5, Eq. (45). Given $b_0\Delta \in \mathfrak{p}$ such that $b_0\Delta \in \mathfrak{f}^\perp$. If $Ak'A' = -b_0\Delta + \bar{c}_0\Delta$, then $\|k'\| \leq h\|b_0\Delta\|$ (killing norm).

$\bar{c}_0 \in \mathfrak{k}$, is bounded $\bar{c}_0 \leq Mhb_0$, where M as before converts between two different norms. Using bounds derived above $b_0' \leq \tilde{M}(Mh + 1)b_0$, and $c_0' \leq \tilde{M}Mhb_0$, $2\tilde{M}(Mh + 1)\Delta<1$, we obtain.

which gives $a_0 + b_0'\Delta + \bar{c}_0 \leq a_0 + b_0\left(\tilde{M}(Mh + 1)\Delta + Mh\right) \leq 1$. For appropriate M', we have

$$\begin{aligned} a_1 &\leq a_0 + \frac{M'}{3}(b_0 + c_0)\Delta \\ b_1 &\leq \frac{M'}{3}(b_0 + c_0)\Delta \\ c_1 &\leq \frac{M'}{3}(b_0 + c_0)\Delta \end{aligned}$$

we obtain

$$a_1 + b_1 + c_1 \leq a_0 + M'(b_0 + c_0)\Delta \leq a_0 + b_0 + c_0$$

where Δ is chosen small.

$$\begin{aligned} U_1 = & \exp\left(-(c_1 + \bar{c}_1)\Delta\right) \exp\left(a_1\Delta + b_1\Delta + c_1\Delta\right) \exp\left(-b_1\Delta + \bar{c}_1\Delta\right) \\ = & \exp\left(-(c_1 + \bar{c}_1)\Delta\right) \exp\left(a_1\Delta + (c_1 + \bar{c}_1)\Delta + b_1'\Delta^2\right) \\ = & \exp\left(a_1\Delta + b_1'\Delta^2 + c_1'\Delta^2\right) \\ = & \exp\left(a_2\Delta + b_2\Delta + c_2\Delta\right) \end{aligned}$$

where $\bar{c}_1 \in \mathfrak{k}$, such that $\bar{c}_1 \leq Mhb_1$.

where, using bounds derived above $b_1' \leq \tilde{M}(Mh + 1)b_1$, and $c_1' \leq \tilde{M}(Mhb_1 + c_1)$, where using the bound $2\tilde{M}(Mh + 1)\Delta < 1$, we obtain

which gives $a_1 + b_1'\Delta + (c_1 + \bar{c}_1) \leq a_1 + ((1 + Mh)b_1 + c_1) \leq a_0 + b_0 + c_0$.

We can decompose, $(b_1' + c_1')\Delta^2$, into subspaces $\left(a_{1'}^{'} + b_2 + c_2\right)\Delta$, where $a_1'' \leq M(b_1' + c_1')\Delta$, $b_2 \leq M(b_1' + c_1')\Delta$ and $c_2 \leq M(b_1' + c_1')\Delta$, where M as before converts between two different norms.

This gives

$$a_2 \leq a_1 + 4\tilde{M}M^2h(b_1 + c_1)\Delta \; b_2 \leq 4\tilde{M}M^2h(b_1 + c_1)\Delta \; c_2 \leq 4\tilde{M}M^2h(b_1 + c_1)\Delta$$

For $x = 8\tilde{M}M^2h\Delta < \frac{2}{3}$, we have, $a_2 + b_2 + c_2 \leq a_1 + (b_1 + c_1) \leq a_0 + b_0 + c_0$,
Using $(b_k + c_k) \leq x(b_{k-1} + c_{k-1}) \leq x^k(b_0 + c_0)$.
Similarly,

$$|a_k - a_{k-1}|_0 \leq x^k(b_0 + c_0)$$

Note, (a_k, b_k, c_k) is a Cauchy sequences which converges to $(a_\infty, 0, 0)$, where

$$|a_\infty - a_0|_0 \leq x(b_0 + c_0) \sum_{k=0}^{\infty} x^k \leq \frac{x(b_0 + c_0)}{1 - x} \leq C\Delta,$$

where $C = 16\tilde{M}M^2h(b_0 + c_0)$.

The above iterative procedure generates k_1' and k_2'' in Eq. (47), such that

$$\exp\left((K_1'Ad_K(X_d)K_1)\Delta\right) = \exp\left(-k_1''\right) \exp\left(a_0\Delta + C\Delta^2\right) \exp\left(-Ak_2''A'\right).$$

where $a_0\Delta + C\Delta^2 \in \mathfrak{f}$. By using a stabilizer H_1, H_2, we can rotate them to $\mathfrak{a}$ such that

$$\exp\left(Ad_K(X_d)\Delta\right)K_1AK_2 = K_aH_1 \exp\left(a_0'\Delta + C'\Delta^2\right)AH_2K_b$$

such that $H_1^{-1}(a_0\Delta + C\Delta^2)H_1 = a_0'\Delta + C'\Delta^2$ is in $\mathfrak{a}$ and $a_0' = P(H_1^{-1}a_0H_1)$ is projection onto $\mathfrak{a}$ such that

$$P(H_1^{-1}a_0H_1) = \sum_k \alpha_k \mathcal{W}_k(X_d).$$

This follows because the orthogonal part of $Ad_{\overline{K}}(X_d)$ to $\mathfrak{f}$ written as $Ad_{\overline{K}}(X_d)^{\perp}$ remains orthogonal of $\mathfrak{f}$

$$\left\langle H^{-1}Ad_K(X_d)^{\perp}H, \mathfrak{a}\right\rangle = \left\langle Ad_K(X_d)^{\perp}, H\mathfrak{a}H^{-1}\right\rangle = \left\langle Ad_K(X_d)^{\perp}, \mathfrak{a}''\right\rangle = 0$$

($a'' \in \mathfrak{f}$), remains orthogonal to $\mathfrak{a}$. Therefore $P(H_1^{-1}a_0H_1) = P\left(H_1^{-1}Ad_{\overline{K}}(X_d)H_1\right) = \sum_k \alpha_k \mathcal{W}_k(X_d)$.

$$\exp\left(Ad_K(X_d)\Delta\right)K_1AK_2 = K_a \exp\left(a + a_{0'}\Delta + C'\Delta^2\right)K_b.$$

Lemma 1 Given $P = K_1 \underbrace{\exp\left(a + a_1\Delta\right)}_{A_1} K_2 = K_3 \underbrace{\exp\left(b - b_1\Delta\right)}_{A_2} K_4$, where $a, b, a_1, b_1 \in \mathfrak{a}$. We can express

$$\exp(b) = K_a \exp\left(a + a_1\Delta + \mathcal{W}(b_1)\Delta\right)K_b,$$

where $\mathcal{W}(b_1)$ is Weyl element of b_1. Furthermore

$$\exp\left(b + b_2\Delta\right) = K_{a''} \exp\left(a + a_1\Delta + \mathcal{W}(b_1)\Delta + \mathcal{W}(b_2)\Delta\right)K_{b''}.$$

Proof. Note, $A_2 = K_3^{-1}PK_4^{-1}$, commutes with b_1. This implies $A_2 = \tilde{K}\exp\left(a + a_1\Delta\right)K$ commutes with b_1. This implies $A_2b_1A_2^{-1} = b_1$, i.e., $\tilde{K}\exp\left(a + a_1\Delta\right)Ad_K(b_1)\exp\left(-(a + a_1\Delta)\right)\tilde{K}' = b_1$, which implies that $Ad_K(b_1) \in \mathfrak{f}$. Recall, from Remark 5,

$$\exp\left(a + a_1\Delta\right)Ad_K(b_1)\exp\left(-(a + a_1\Delta)\right) = \sum_k c_k\left(Y_k \cos(\lambda_k) + X_k \sin(\lambda_k)\right).$$

This implies $\sum_k c_k \sin(\lambda_k)X_k = 0$, implying $\lambda_k = n\pi$. Therefore,

$$\exp\left(2(a + a_1\Delta)\right)Ad_K(b_1)\exp\left(-2(a + a_1\Delta)\right) = Ad_K(b_1).$$

We have shown existence of H_1 such that $H_1Ad_K(b_1)H_1^{-1} \in \mathfrak{a}$, using H_1, H_2 as before,

$$\begin{aligned}\tilde{K}\exp\left(a + a_1\Delta\right)K\exp\left(b_1\Delta\right) &= \tilde{K}H_2\exp\left(a + a_1\Delta\right)H_1\exp\left(Ad_K(b_1)\Delta\right)K \\ &= K_a\exp\left(a + a_1\Delta + \mathcal{W}(b_1)\Delta\right)K_b.\end{aligned}$$

Applying the theorem again to

$$K_a\exp\left(a + a_1\Delta + \mathcal{W}(b_1)\Delta\right)K_b\exp\left(b_2\Delta\right) = K_{a''}\exp\left(a + a_1\Delta + \mathcal{W}(b_1)\Delta + \mathcal{W}(b_2)\Delta\right)K_{b''}.$$

Lemma 2 Given $P_i = K_1^iA^iK_2^i = K_1^i\exp\left(a^i\right)K_2^i$, we have $P_{i,i+1} = \exp\left(H_i^+\Delta_i^+\right)P_i$, and $P_{i,i+1} = \exp\left(-H_{i+1}^-\Delta_{i+1}^-\right)P_{i+1}$, where $H_i^+ = Ad_{K_i}(X_d)$. From above we can express

$$P_{i,i+1} = K_a^{i+}\exp\left(a^i + a_1^{i+}\Delta_+^i + a_2^{i+}\left(\Delta_+^i\right)^2\right)K_b^{i+}.$$

where a_1^{i+} and a_2^{i+} are first and second order increments to a_i in the positive direction. The remaining notation is self-explanatory.

$$P_{i,i+1} = K_a^{(i+1)-} \exp\left(a^{i+1} - a_1^{(i+1)-}\Delta_-^{i+1} - a_2^{(i+1)-}\left(\Delta_-^{i+1}\right)^2\right) K_b^{(i+1)-}.$$

$$\exp\left(a^{i+1}\right) = K_1 \exp\left(a^i + a_1^{i+}\Delta_+^i + a_2^{i+}\left(\Delta_+^i\right)^2 + \mathcal{W}\left(a_1^{(i+1)-}\Delta_-^{i+1} + a_2^{(i+1)-}\left(\Delta_-^{i+1}\right)^2\right)\right) K_2.$$

$$\mathcal{W}\left(a_1^{(i+1)-}\Delta_-^{i+1} + a_2^{(i+1)-}\left(\Delta_-^{i+1}\right)^2\right) = \mathcal{P}\left(\mathcal{W}\left(a_1^{(i+1)-}\right)\right)\Delta_-^{i+1} + \mathcal{P}\left(\mathcal{W}\left(a_2^{(i+1)-}\right)\right)\left(\Delta_-^{i+1}\right)^2$$

$$= \sum_k \alpha_k \mathcal{W}_k(X_d)\Delta_-^{i+1} + o\left(\left(\Delta_-^{i+1}\right)^2\right)$$

where, $a^i, a_1^i, a_2^i \in \mathfrak{a}$.
Using Lemma 1 and 2, we can express

$$P_n(T) = K_1 \exp\left(a_n\right) \exp K_2 = K_1 \exp\left(\sum_i \mathcal{W}(a_i^+)\Delta_i^+ + \mathcal{W}(a_{i+1}^-)\Delta_{i+1}^-\right) \exp\left(\underbrace{\sum o(\Delta^2)}_{\leq \varepsilon T}\right) K_2$$

Letting ε go to 0, we have

$$P_n(T) = K_1 \exp\left(T \sum_i \alpha_i \mathcal{W}_i(X_d)\right) K_2.$$

Hence the proof of theorem. **q.e.d.**

4. Conclusion

In this chapter, we studied some control problems that derive from time optimal control of coupled spin dynamics in NMR spectroscopy and quantum information and computation. We saw how dynamics was decomposed into fast generators $\mathfrak{k}$ (local Hamiltonians) and slow generators $\mathfrak{p}$ (couplings) as a Cartan decomposition $\mathfrak{g} = \mathfrak{p} \oplus \mathfrak{k}$. Using this decomposition, we used some convexity ideas to completely characterize the reachable set and time optimal control for these problems.

Author details

Navin Khaneja
Systems and Control Engineering, IIT Bombay, India

*Address all correspondence to: navinkhaneja@gmail.com

References

[1] Nielsen M, Chuang I. Quantum Information and Computation. New York: Cambridge University Press; 2000

[2] Ernst RR, Bodenhausen G, Wokaun A. Principles of Nuclear Magnetic Resonance in One and Two Dimensions. Oxford: Clarendon Press; 1987

[3] Cavanagh J, Fairbrother WJ, Palmer AG, Skelton NJ. Protein NMR spectroscopy. In: Principles and Practice. New York: Academic Press; 1996

[4] Khaneja N, Brockett RW, Glaser SJ. Time optimal control of spin systems. Physical Review A. 2001;**63**:032308

[5] Khaneja N, Glaser SJ. Cartan decomposition of $SU(2^n)$ and control of spin systems. Chemical Physics. 2001; **267**:11-23

[6] D'Alessandro D. Constructive controllability of one and two spin 1/2 particles. Proceedings 2001 American Control Conference; Arlington, Virginia; June 2001

[7] Kraus B, Cirac JI. Optimal creation of entanglement using a two qubit gate. Physical Review A. 2001;**63**:062309

[8] Bennett CH, Cirac JI, Leifer MS, Leung DW, Linden N, Popescu S, et al. Optimal simulation of two-qubit hamiltonians using general local operations. Physical Review A. 2002;**66**: 012305

[9] Khaneja N, Glaser SJ, Brockett RW. Sub-Riemannian geometry and optimal control of three spin systems. Physical Review A. 2002;**65**:032301

[10] Vidal G, Hammerer K, Cirac JI. Interaction cost of nonlocal gates. Physical Review Letters. 2002;**88**:237902

[11] Hammerer K, Vidal G, Cirac JI. Characterization of nonlocal gates. Physical Review A. 2002;**66**:062321

[12] Yuan H, Khaneja N. Time optimal control of coupled qubits under non-stationary interactions. Physical Review A. 2005;**72**:040301(R)

[13] Yuan H, Khaneja N. Reachable set of bilinear control systems under time varying drift. System and Control Letters. 2006;**55**:501

[14] Zeier R, Yuan H, Khaneja N. Time optimal synthesis of unitary transformations in fast and slow qubit system. Physical Review A. 2008;**77**: 032332

[15] Yuan H, Zeier R, Khaneja N, Lloyd S. Constructing two qubit gates with minimal couplings. Physical Review A. 2009;**79**:042309

[16] Reiss T, Khaneja N, Glaser S. Broadband geodesic pulses for three spin systems: Time-optimal realization of effective trilinear coupling terms and indirect SWAP gates. Journal of Magnetic Resonance. 2003;**165**:95

[17] Khaneja N, Glaser S. Efficient transfer of coherence through Ising spin chains. Physical Review A. 2002;**66**: 060301

[18] Khaneja N, Heitmann B, Spörl A, Yuan H, Schulte-Herbrüggen T, Glaser SJ. Shortest paths for efficient control of indirectly coupled qubits. Physical Review A. 2007;**75**:012322

[19] Yuan H, Zeier R, Khaneja N. Elliptic functions and efficient control of Ising spin chains with unequal couplings. Physical Review A;**77**:032340

[20] Yuan H, Khaneja N. Efficient synthesis of quantum gates on a three-

spin system with triangle topology. Physical Review A;**84**:062301

[21] Yuan H, Wei D, Zhang Y, Glaser S, Khaneja N. Efficient synthesis of quantum gates on indirectly coupled spins. Physical Review A;**89**:042315

[22] Redfield AG. The theory of relaxation processes. Advances in Magnetic and Optical Resonance. 1965; **1**:1-32

[23] Lindblad G. On the generators of quantum dynamical semigroups. Communications in Mathematical Physics. 1976;**48**:199

[24] Helgason S. Differential Geometry, Lie Groups, and Symmetric Spaces. Cambridge: Academic Press; 1978

[25] Brockett RW. System theory on group manifolds and Coset spaces. SIAM Journal of Control. 1972;**10**: 265-284

[26] Jurdjevic V, Sussmann H. Control systems on lie groups. Journal of Differential Equations. 1972;**12**:313-329

[27] Jurdjevic V. Geometric Control Theorey. New York: Cambridge University Press; 1997

[28] Kostant B. On convexity, the Weyl group and the Iwasawa decomposition. Annales scientifiques de l'cole Normale SupÃ©rieure. 1973;**6**(4):413-455